T0076524

Women in Engineering and Science

Series Editor

Jill S. Tietjen
Greenwood Village, Colorado, USA

More information about this series at http://www.springer.com/series/15424

Alice Cline Parker • Leda Lunardi
Editors

Women in Microelectronics

Editors
Alice Cline Parker
Department of Electrical and Computer Engineering
University of Southern California
Los Angeles, CA, USA

Leda Lunardi
Department of Electrical and Computer Engineering
North Carolina State University
Raleigh, NC, USA

ISSN 2509-6427 ISSN 2509-6435 (electronic)
Women in Engineering and Science
ISBN 978-3-030-46893-4 ISBN 978-3-030-46377-9 (eBook)
https://doi.org/10.1007/978-3-030-46377-9

This Springer imprint is published by the registered company Springer Nature Switzerland AG
The registered company address is: Gewerbestrasse 11, 6330 Cham, Switzerland

"To our children"

Foreword

This unique book captures the stories and perspectives of pioneering women in microelectronics, providing insight into the technical and personal challenges they faced, and how they were able to overcome these to succeed. It is our hope that these role models will encourage and inspire the next generation of bold, innovative engineers to shape a brighter future for all.

Dr. France Córdova, former Director of the U.S. National Science Foundation, wrote "Diversity—of thought, perspective, and experience—is essential for excellence in research and innovation in science and engineering." This is because diverse teams comprise a wider range of viewpoints and skill sets, i.e., greater collective intelligence. Women should benefit from engineering innovations as much as men. For these reasons, it is important for women to be included in engineering teams to develop better solutions and improve the quality of life for all members of our society.

Engineers continually innovate new processes, devices, and systems to solve problems and improve the quality of life. Perhaps the most prominent example of this is the steady advancement of microelectronics technology over the past 60 years that sustained an exponential pace of performance improvement in information processing, communication, and storage. Indeed, the information technology revolution has had a transformative impact on every aspect of life in modern society.

The confluence of cloud computing, big data, artificial intelligence, and the Internet of Things today is projected to give rise to digital transformation in all sectors of the economy. This vision of the future cannot be realized without new innovations, however, due to the physical limits of transistor miniaturization and fundamental energy efficiency limits of digital computing. It is widely recognized that new electronic materials, manufacturing processes, logic and memory devices, integrated circuit chip architectures, etc. will be needed to circumvent these limitations.

Dean and Roy W. Carlson Professor
of Engineering, College of Engineering
University of California
Berkeley, USA

Tsu-Jae King Liu

Contents

Chapter 1
Introduction

Alice Cline Parker and Leda Lunardi

1.1 Scope

Electrical engineering was essentially born in the twentieth century, with women involved from the beginning, if you set aside Ben Franklin's experiments with kites, lightning rods, and electricity two centuries earlier, and observations about charge and electricity dating back to the ancient Greeks. Edith Clarke, considered the first professional woman electrical engineer, analyzed electrical power systems and Margaret Mary Partridge, also an electrical engineer, began involving and supporting women in engineering. Women entered the engineering profession in unorthodox ways, exemplified by actress Hedy Lamarr, who made major contributions to communication theory during World War II. Kitty O'Brien Joyner began as an electrical engineer, but made major contributions to aeronautics and flight during her long career with NASA. Another electrical engineer, Thelma Estrin, had a long and distinguished career in the field of biomedical engineering.

The first point-contact transistor was invented at Bell Labs in 1947, and from that humble beginning of the golden age in microelectronics to what has been an engineering revolution, women have been involved. Now, we open car doors,

A. C. Parker (✉)
Department of Electrical and Computer Engineering, University of Southern California,
Los Angeles, CA, USA
e-mail: parker@usc.edu

L. Lunardi
Department of Electrical and Computer Engineering, North Carolina State University,
Raleigh, NC, USA
e-mail: leda_lunardi@ncsu.edu

A. C. Parker, L. Lunardi (eds.), *Women in Microelectronics*, Women
in Engineering and Science, https://doi.org/10.1007/978-3-030-46377-9_1

communicate, preserve virtually every moment of our lives, and know who is at the front door when we are thousands of miles away. Very few women were part of the microelectronics revolution that has moved us from the industrial age to the information age. As more women are entering the microelectronics field, it is important to preserve the history of early pioneers, to understand and celebrate the challenges these women have overcome, to inspire young women, and to raise awareness of contributions.

Mildred Dresselhaus, Institute Professor at Massachusetts Institute of Technology, inspired electrical engineers with her accomplishments on electronic properties of materials. The VLSI (very large-scale integration) revolution that provided a design approach to today's integrated circuit chips was led by Lynn Conway, along with Carver Mead, California Institute of Technology Professor. Mead's former doctoral student, Misha Mahowald, along with Mead, triggered the current wave of activity in neuromorphic computing, modeling the brain with electronics.

What is the combination of traits and circumstances that motivated these women to succeed in engineering? We, the editors, have assembled some chapters describing women's journeys and career paths related to microelectronics and beyond. Along with documenting what it took to succeed, we hope that the paths followed will inspire young women in this field, and give them courage to persist and succeed. This book contains stories of women engineers' paths through the golden age of microelectronics.

These stories, like the biographies of Marie Curie and the National Geographic's stories of Jane Goodall's research that inspired the authors, will inspire and guide young women along unconventional pathways to contributions that we can only begin to imagine. The reasons and how the women contributors here chose their career paths and how they navigated their careers will be of interest to anyone, from university advisors to industry Chief Executive Officers, who can imagine the advantages of a future with a diverse workforce. One word that comes to mind when editing this book is "unconventional."

The book will help managers and industry leaders to understand the environment these women not only survived, and persisted, but also blossomed in, to enlighten future decision makers. It will inspire young women to pursue unconventional careers. The stories in this book are timeless. A century from now, the chapters will contain stories told by some brave pioneers at the beginnings of a new age.

1.2 Chapter Introductions by Author

Susan Palmateer: "From Crystal Growth to New Product Introduction"

Only in the late 1970s, compound semiconductor epitaxy technologies were developed enabling the precise control of the different doping levels and the thickness of each layer, "band gap engineering". With more advances in the bulk crystal growth and diverse epitaxial growth techniques, it rapidly expanded the field for compound semiconductor applications. In the field of communications, electronic devices with wider bandwidth and lower noise were one of them: radio transmission and detection including radar equipment as well as commercial systems.

Dr. Palmateer, who started as a compound semiconductor materials grower moving through different paths, tells her journey on becoming a group leader in industry and what it took.

Adrienne Stiff-Roberts: "Thin-Film Deposition of Hybrid Materials"

Conventional semiconductors (silicon, germanium) are commonly found in nature as crystals. Compound semiconductors like gallium arsenide (GaAs) or indium phosphide (InP) are "manufactured" in controlled conditions in a laboratory. New techniques have been investigated to create hybrid materials comprised of organic and inorganic components primarily for optoelectronic device applications, such as infrared photodetectors, photovoltaic solar cells, light-emitting diodes, and sensors.

Prof. Stiff-Roberts presents her research on resonant infrared matrix-assisted pulsed laser evaporation for the thin-film deposition of hybrid materials. One of the challenges of the research is to control the co-deposition of two or more materials with different chemical and physical properties in order that the synthesis has pre-determined properties and functionality.

Missing Santosh Kurinec.

Santosh Kurinec: "Nanoscale Materials Engineering for Microelectronics"

Prof. Kurinec, a multidisciplinary expert on magnetic, optical, and electronic materials, tells her journey how to conduct a world class research on integrating novel materials and devices onto CMOS platform for heterogeneous integration. With her students, she required inter-disciplinary, inter-institutional, and global collaborations.

Zeynep Çelik-Butler: "Smart Skin: Multifunctional, Flexible, Conformable Sensor Arrays"

The evolution of electronic devices and their integration enabled the development of applications that required to conform to different surfaces. Large areas and flexible electronics have been researched for data processing, specifically in the areas of decision making, biochemical sensing, and energy harvesting.

Prof. Çelik-Butler reviews her career in a chapter dedicated to the topic of Smart Skin, a term defined to mimic the human skin, that is extensively applied to robotics.

Xiuling Li: "Nanodevices and Applications: My Nonlinear Career Trajectory"

An important aspect in microelectronic devices is the course taken from the concept, the device structure and fabrication development. The magic behind different approaches and processing steps is a multidisciplinary task, requiring critical decisions that lead to further integration and applications.

Through her career in industry and academia, Dr. Li has processed 3D structures and devices with bottom-up growth and top-down fabrication that led to numerous applications in the disciplines of electronics, photonics, nanotechnology, and quantum technology.

Ru Huang: "Gate-All-Around Silicon Nanowire Transistor Technology"

Prof. Huang has been exploring the fundamental limits of scaling field effect transistors by investigating vertical nanowire transistors with all-around gate geometrics for an optimum field effect control of the channel carriers.

Dr. Huang presents her chapter on silicon-based transistors research including the gate-all-around design to enhance the conductivity while easing the ohmic contact formation.

Leda Lunardi: "Heterojunction Bipolar Transistors and Monolithically Integrated Photoreceivers Among Other Applications"

Incorporation of a wide bandgap emitter was proposed by Schockley in 1951 and Kroemer in 1957; however, it took almost two decades in the development of optimum growth conditions and epitaxial material technology for compound semiconductors to unleash the potential of heterojunction bipolar transistors.

Prof. Lunardi reviews her contributions on two compound semiconductors heterojunction bipolar transistor technologies: GaAs and InP, specifically in the development of photoreceivers for optical communication applications.

Mona Zhagoul: "Integrated Circuits, MEMS, and Nano Electronics for Sensors Applications"

Prof. Zhagoul started her career as a circuit designer. As soon as she had an opportunity to learn CMOS, she employed it in analog and digital applications. Over the years, she has incorporated sensors in her circuits using PZT as the piezoelectric material, and temperature sensors using thermocouple techniques developing chemical and gas sensors. By using microelectromechanical systems (MEMS) and nanoelectromechanical systems (NEMS), she has been able to expand the range of applications to power sensors and micromachined accelerometers. Her more recent interest includes synthesis of 2D materials and their inclusion in the design of electronic circuits.

Rhonda Franklin: "From Microwave Communication Systems to Nanomedicine Tools: Using Advanced Microelectronics Fabrication as an Enabler"

Prof. Franklin started her career in the microwave space communications with focus on interconnect technology. She developed nanofabrication methods for circuits and devices, and characterization techniques in order to understand electric/magnetic materials. She developed techniques for reducing electric field and crosstalk on transmission lines and other passive components at the wafer-level of unpackaged circuits.

Dr. Franklin employs nanofabrication techniques into cells as biomarkers for cancer research.

Jennifer Hasler: "Becoming a Creative Female Engineer … Enabling the Next Computing Opportunities"

Prof. Hasler tells her creative and personal story, starting and becoming an electrical engineer. She recalls the details of discovery on her research trajectory and working on the cutting edge of computation and neuromorphic techniques, and then becoming a female engineer.

Alice Parker: "From Silicon to the Brain Using Microelectronics as a Bridge"

The research journey of Alice Cline Parker is presented in this chapter. Alice Parker's family background and the circumstances that propelled her into engineering begin the chapter. The journey her research took from a more-conventional electrical engineer into the neuroscience-inspired world of neuromorphic circuits completes the chapter. The role of nanotechnology in her research directions is also highlighted. Parker cites the individuals who were key to her career success, beginning with her father's influence.

Telle Whitney; "VLSI and Beyond: The Dream of Impact, Creating Technology with Inclusive Cultures"

Telle Whitney describes her path from her early days in Salt Lake City to becoming CEO of the Anita Borg Institute, creating and leading the organization to one of global impact, with a reach of over 750 K technical women. In the field of VLSI design, Telle was a trailblazer, both in the early days of the VLSI revolution and as an entrepreneur in the semiconductor field. She is a recognized leader both as a semiconductor pioneer and as award-winning leader in the movement for women creating technology that impacts everyone's life.

Gabriele Saucier: "Creating Collaborative Design Environments"

This chapter relates the journey of a girl born in a remote European region under war and who had little chance for accessing engineering study, but who obstinately wanted to learn, understand the world around her, and "participate".

From PHD to a university professor position, from logic design to real circuit layout and ultimately to web-based silicon IP exchange, she met obstacles as a mother and a woman but also got support from the IEEE community, from EU projects, and nice people that she had a chance to meet in her life.

> I wish to all women of the world to be as obstinate as I was (and even more) to learn, understand and participate in their world which may not be an easy one …

Diana Marculescu: "The Quest for Energy Aware Computing: Confessions of an Accidental Engineer"

Dr. Marculescu's contributions to microelectronics started with electronic design automation tools. With her computer science skills and expanding emerging platforms, she has incorporated into design parameters related to power consumption, reliability, and the nanoscale semiconductor.

Lauren Palmateer: "Semiconductors to Light Antennas: A Woman Engineer's Career at the Turn of the Third Millennium"

Starting with an earlier technical internship in cosmology and following her sister's path to engineering school, Dr. Palmateer completed her studies with a graduate degree on III–V heterojunction field effect transistors. After graduate school, her industrial career continued in different areas including thin-film transistors and reflective displays.

In her chapter, Dr. Lauren Palmateer gives a candid account of her continuous learning journey as a world traveler.

1.3 Epilogue: Changes and Rising Stars

The last decade has seen a blossoming of interest in careers in engineering, science, and technology, particularly for women. Notable in the maker space is AdaFruit Industries founder Limor Fried. Jeri Ellsworth has blazed trails as an electronic gaming entrepreneur. Entrepreneur Courtney Gras is inspiring student entrepreneurship, and fellow electrical engineer Christina Koch is an astronaut. Now the women of electrical engineering are too numerous to mention, even in the microelectronics domain. Recognition by professional societies and other organizations has highlighted many of these women following in the footsteps of the pioneers included here.

At the time of this book printing, 50% of freshmen at USC in the Viterbi School of Engineering and 50% of freshmen in the College of Engineering at NC State are women. We are looking forward to the coming decades when women in microelectronics will be the norm, not the exception. As you explore the chapters that follow, note the exceptional circumstances and persistence that led these women to careers in microelectronics. If you are an academician or industry leader supporting the advancement of women in engineering, these chapters might give a better understanding of the challenges faced. If you are an aspiring engineer, the editors hope you will find inspiration, support, and encouragement from reading these engineers' journeys.

Chapter 2
From Crystal Growth to New Product Introduction

Susan Palmateer

2.1 The Impact of Diversity, Mentoring, Networking, and Teamwork on My Leadership Journey

I was born on July 2, 1957, in New Jersey to Helen L. Palmateer and George R. Palmateer. I spent my early years in Fort Lee, New Jersey, and I remember Palisades Amusement Park where I enjoyed many weekends with my family. I moved to Norwood, New Jersey, when I was 5 years old with my family and younger sister Lauren Fay Palmateer—another engineer with a Ph.D. in electrical engineering from Cornell University.

Now that's interesting—two female engineers with Ph.D.s!

How did we both end up with Ph.D.s from Cornell? We had great mentors and encouragement from our parents. At a young age, teachers like Mr. Skorka, Ms. Besold, and Ms. Sovolus saw our potential in math and science and brought it out. In high school, our role models and mentors were great too—Mr. Kimball, Ms. Kelly, Mr. Wolfe, and Mr. Donofrio. The diversity and inclusion was a strong influencer growing up in New Jersey outside of Manhattan where Italians, Polish, Jews, and blue-collar and white-collar workers all came together. Did forget to mention grades kindergarten through fifth? We were in Catholic school, and while not great with academics, coaching, or mentoring, it instilled the benefit of hard work.

Therefore, off I went to Monmouth College in 1975. I had money for the first year, which was $5000 plus scholarships. In my senior year at Monmouth, I got an internship at Bell Laboratories. It was exciting times in technology—lots of funding and some of the most renowned scientists. I had an opportunity to learn from Dr. Martin Schneider and Dr. Cho, the father of molecular beam epitaxy. I would later be known as the mother of molecular beam epitaxy.

S. Palmateer (✉)
BAE Systems, Nashua, NH, USA
e-mail: susan.palmateer@baesystems.com

A. C. Parker, L. Lunardi (eds.), *Women in Microelectronics*, Women in Engineering and Science, https://doi.org/10.1007/978-3-030-46377-9_2

Now, how did my sister Lauren follow the engineering path? A little story: one day she came to visit me at Bell Labs, where she met Dr. Arno Penzias, a Nobel Prize winner at Crawford Hill where he and Dr. Robert Carlson were collecting background radiation with an antenna. Dr. Penzias came to me a few days later and asked me if my sister would be interested in a job working for him. He told me she asked him a question he was not able to answer and said, "I want to hire that woman!"

Lauren transferred to Monmouth College from SUNY Albany and got a bachelor's degree in electrical engineering while running a telescope and collecting data for Dr. Penzias and Dr. Carlson for the next 3 years. I assume this data is what contributed to their winning a Nobel Prize for the big bang theory.

We then both went on to earn Ph.D.s from Cornell because of our mentor, Dr. Martin Schneider at Bell Laboratories. In the interim, Lauren worked for IBM as well as other large and small companies. (Read more in Lauren Palmateer's chapter of this book.)

So, how did we both end up with PhDs in electrical engineering? It was the impact of diversity, mentoring, coaching, and networks we developed. It started with a mother who said, "You can do and be whatever you want." She encouraged us to be independent and pursue careers so we could support ourselves and not depend on someone else. Figure 2.1 captures my leadership journey timeline with the influence of teamwork, adaptability, attitude, and diversity, overlaid on my academic and work experiences. I constructed this to talk to the next generation of women entering the engineering field.

What did I see and learn from my mentors? Drive and passion. Professor Lester Eastman, one of my graduate degree professors, was my most influential mentor. He had drive and passion, and saw no difference between men or women or minorities. He treated everyone equally. The greatest advice Professor Eastman passed on to me, and all 125 of his graduate students, was to work as a team on technology that made a difference in performance, and develop technology that could transition into products.

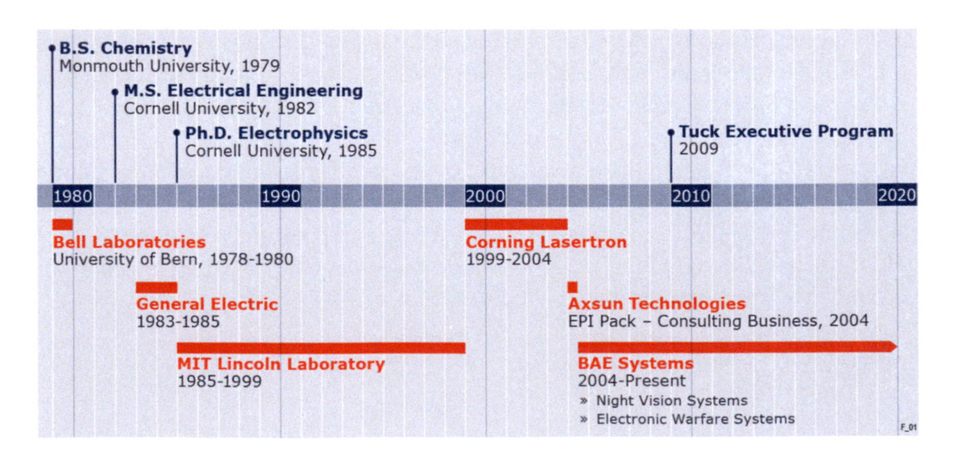

Fig. 2.1 Dr. Palmateer's leadership journey timeline

2.2 Life Challenges and Insights

I faced a lot of adversity as a female engineering student, and early in my career.

I was approached several times with suggestive implications by influential and successful male engineers and leaders, and politely said, "No, thank you." I was told that I'm "not as smart as the guys" by a professor and an advisor. I was also told I was "not as capable" as the men and that I "shouldn't be here" by an advisor. I was hit on the buttocks by a research advisor, which left a hand print for 7 days, and was then told by colleagues that I "asked for it." This was my most challenging lesson in life. I had a male colleague move all my chairs and various other lab equipment into the hallway each day to slow me down, and another say, "She won't work again—she's pregnant." Not to mention the sexually suggestive items in the workplace in the 1980s.

While I don't feel any woman should have to experience these challenges on their path to becoming an engineer, the reality is that we are often doubted and pushed outside of our comfort zones. I faced these situations with a positive attitude and used each one as an opportunity to advocate for myself and other women, in the process of learning and teaching many valuable lessons about perseverance. No one was going to get in the way of my dreams to earn a Ph.D. I tell you this so that you feel empowered to do the right thing, and be resilient in the face of adversity. Future generations of young women are counting on you to pave the way and be mentors. Together we can work to remove barriers preventing them from reaching their goals.

2.3 Crystal Growth—Molecular Beam Epitaxy, Organometallic Chemical Vapor Deposition, and Gas Source Molecular Beam Epitaxy

2.3.1 Molecular Beam Epitaxy

I was first introduced to molecular beam epitaxy (MBE) and III–V semiconductors at Bell Laboratories in the optical devices group in Holmdel, New Jersey, where I learned from the experts. From Bell Labs, I went off to Cornell where I did both my master's and Ph.D. [1, 2] research in MBE and the diffusion and impurity incorporation in III–V semiconductors. The out diffusion and redistribution of substrate related impurities during MBE growth was shown to adversely affect device performance. An understanding of the diffusion mechanisms in substrates and epitaxial layers was developed to devise a way of controlling atomic movement of impurities during MBE growth. The redistribution of impurities was found to be influenced by surface electric fields and vacancies in the semiconductor [3, 4] (Fig. 2.2).

Planar doped barrier structures were shown to be extremely sensitive to MBE growth conditions specifically intrinsic doping levels (i.e., unintentional doping levels and deep levels), which can be substrate dependent. Heat treatment of Gallium

Fig. 2.2 Far above Cayuga's Waters—Cornell University, Ithaca, NY

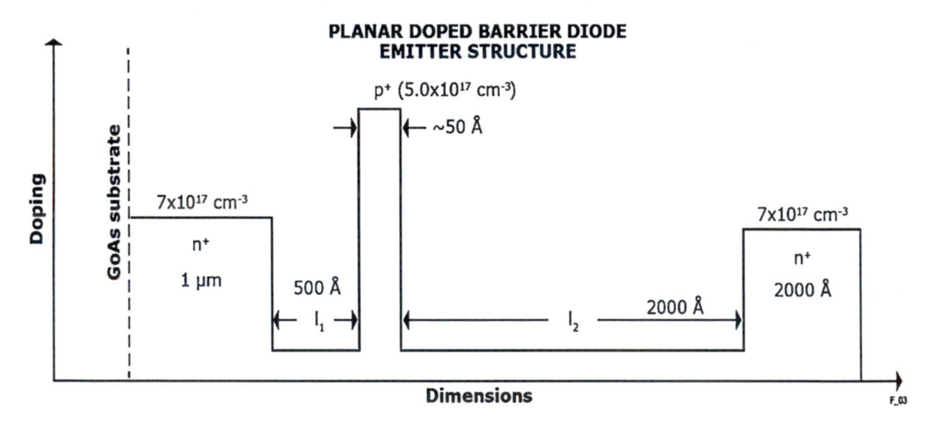

Fig. 2.3 Planar doped barrier structure [5]

Arsenide (GaAs) substrates in combination with controlling the growth conditions resulted in precise control of designed barrier heights (Fig. 2.3).

I finished my Ph.D. research at General Electric in Syracuse, New York. I was in charge of producing MBE material for high frequency and low-noise microwave devices. By pre-annealing substrates and carefully controlling the V:III flux ratio during growth, extremely high-purity GaAs and GaAs/Aluminum Gallium Arsenide (AlGaAs) quantum wells were grown. In 1985, I produced the highest mobility GaAs as shown in Fig. 2.4 and single quantum well structures as shown in Fig. 2.5 by MBE.

Critical to these successes was careful startup of the MBE system combined with understanding the growth thermodynamics-As4 flux ratio, growth temperature,

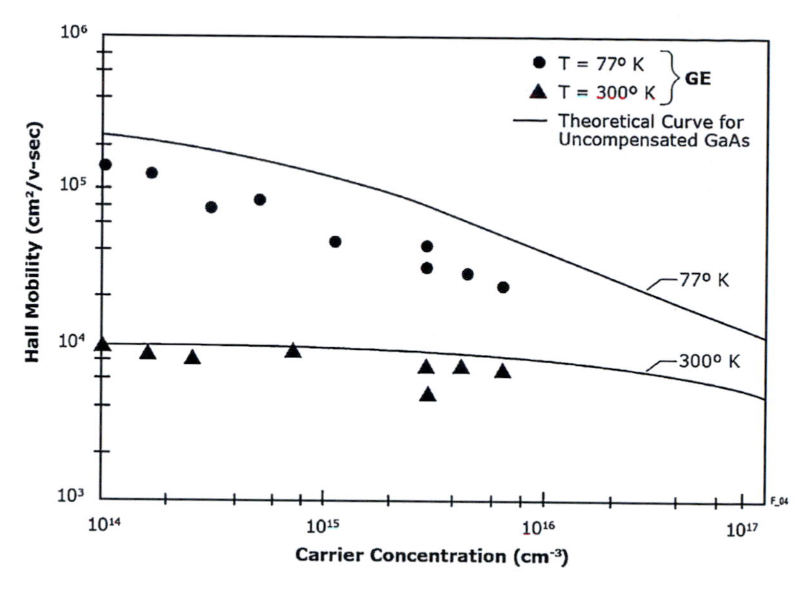

Fig. 2.4 Plot of theoretical and experimental 77K Mobility verses carrier concentration for n-type GaAs [2]

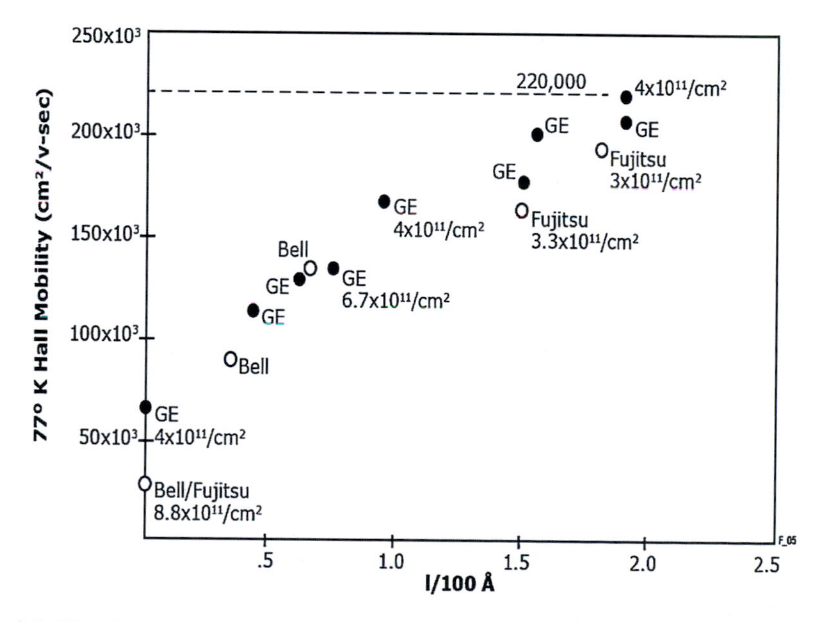

Fig. 2.5 Plot of 77K mobility versus l/100 (l = spacer layer thickness) [2]

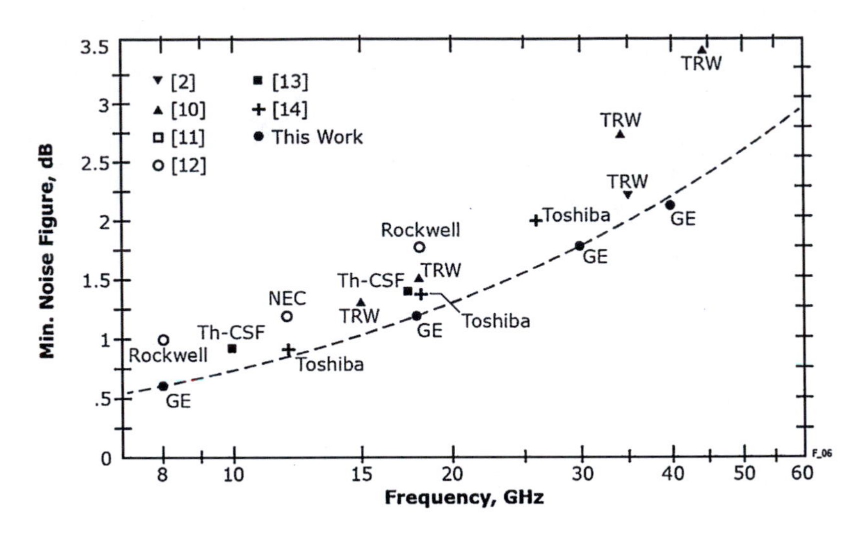

Fig. 2.6 Reported HEMT noise performance at room temperature from major laboratories [6]

impurity redistribution and thermal conversion of the growth front. My manager and mentor, Dr. James Hwang, provided the resources and coaching to successfully create this capability.

This high-purity material resulted in the lowest noise high-electron-mobility transistors ever produced with the lowest noise figure to date demonstrated in 1985 as shown in Fig. 2.6.

It was during this time at General Electric that I first developed an intense interest in process control and process variability, and my passion for transitioning technology into products.

Accomplishments and Insights
I produced the highest purity MBE material to date and set up a manufacturing infrastructure for microwave devices. What enabled me to do this was chemistry training (attention to details) and mentorship by Dr. Jim Hwang, who provided the funding, which fueled my drive and passion.

2.3.2 Organometallic Chemical Vapor Deposition

In the late 1980s at MIT Lincoln Laboratories, we (Dr. Christine Wang, Dr. Steve Groves, James Caunt, Dave Hovey, and myself) designed and built two new organometallic chemical vapor deposition (OMCVD) reactors for the growth of compound semiconductors GaAs/AlGaAs and Gallium Indium Arsenide Phosphide/Indium Phosphide (GaInAsP/InP). With gas flow visualization, tracer gas studies, and numerical modeling, these reactor designs were optimized for highly uniform, reproducible, abrupt interface growth. High-performance laser diodes were produced in each reactor for each respective material system [7].

It was shown that convective cells might form when a gas of a different density than that of the gas already flowing in the tube is rapidly switched into the reactor, even when the run-vent lines of the switching manifold are pressure- and flow-balanced. This density effect was reported for the first time, and was found to severely degrade the material quality of InP/InGaAs and Indium Phosphide/Indium Gallium Phosphide (InP/InGaP) heterojunctions.

The above shown in Fig. 2.7 is referred to as a stagnation flow reactor designed and developed by Dr. Christine Wang and James Caunt. Tangential gas injection resulted in fully developed flow that could be made laminar by disk rotation. Gas injection through a porous plug produced the most uniform flow to the disk. In addition, gas residence times were minimized when this injection geometry was combined with low pressure. Extremely uniform GaAs and AlGaAs epilayers were grown in this vertical rotating-disk OMCVD reactor. Diode lasers were produced with threshold current densities, differential quantum efficiencies, and emission wavelengths that were highly uniform over 16 cm^2. Excellent wafer-to-wafer reproducibility of the emission wavelengths was demonstrated [8].

Figure 2.8 shows sulfur hexafluoride (SF_6) detected at the center of the susceptor in the stagnation flow reactor as a function of time after gas switching for different density changes. These tracer gas experiments were instrumental in designing OMCVD reactors capable of producing abrupt interfaces with excellent across wafer uniformity.

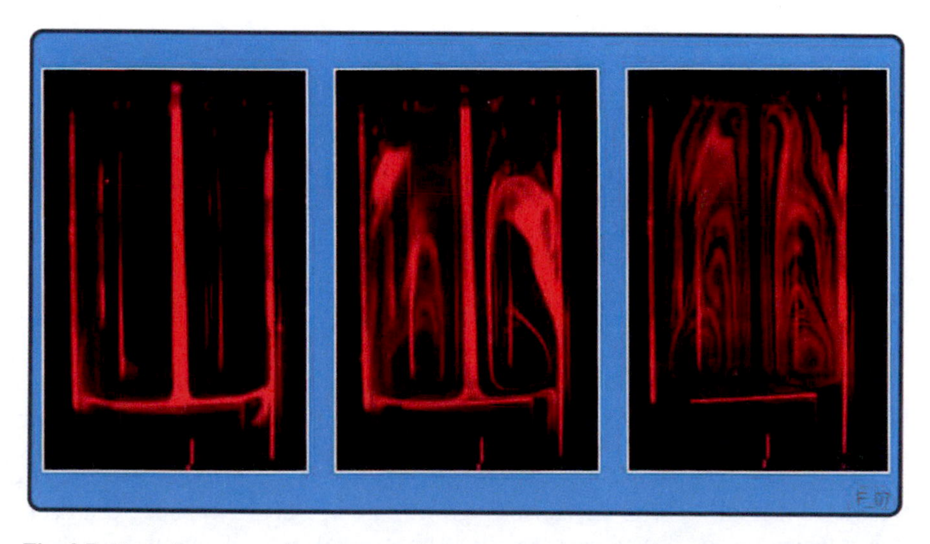

Fig. 2.7 Vertical cross sections showing flow patterns at room temperature and 1 atmosphere obtained for gas injection through a coaxial pipe inlet: (**a**) 1 s after initial introduction of smoke; (**b**) 20 s after initial introduction of smoke; (**c**) 1 min after smoke was turned off. Intense gas recirculation occurs throughout the reactor as a result of the impinging jet. "Provided by MIT Lincoln Laboratory" [8]

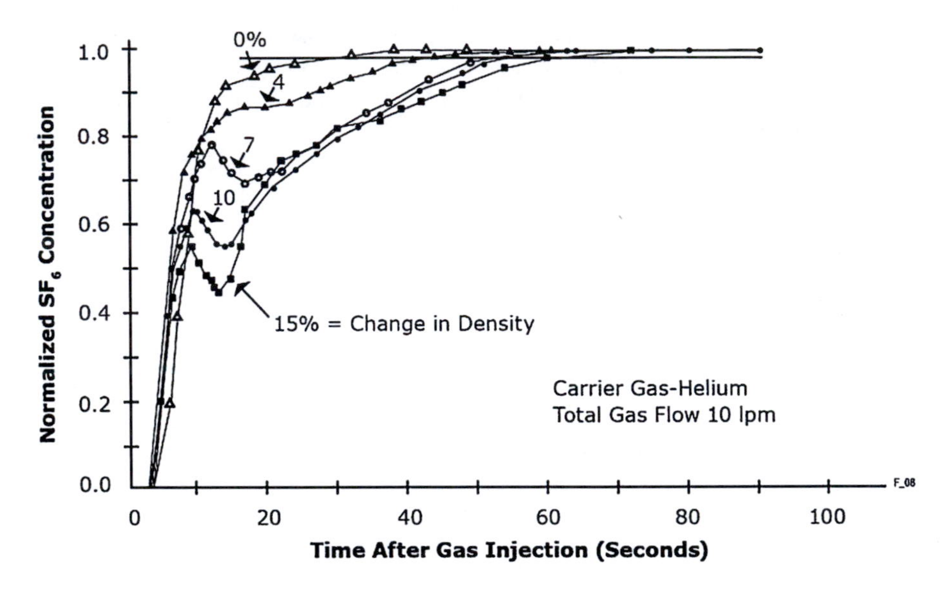

Fig. 2.8 Normalized SF_6 concentration versus time after gas injection. The total flow through the reactor is 10 L/min of helium. Increasing amounts of nitrogen are switched into the gas stream to study the effect of increasing density changes [9]

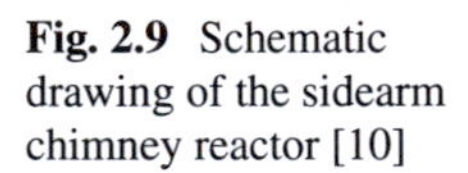

Fig. 2.9 Schematic drawing of the sidearm chimney reactor [10]

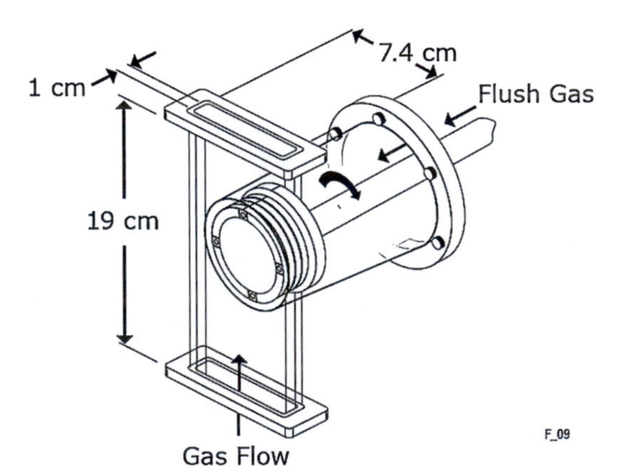

The "chimney" reactor design is shown in Fig. 2.9 was designed and developed in collaboration with Steve Groves and James Caunt. This reactor was optimized for the growth of InGaAsP related materials. By decreasing the cross-sectional area and increasing the gas velocity through the tube, the formation of convective cells is restricted. For the case of InP/InGaAs growth, a large density change can occur when the arsine and phosphine are switched because AsH3 is about twice as dense as PH3. The most obvious way to eliminate convection due to density changes is to density balance the gases switched between the reactor and vent line.

InGaAs strained-layer quantum well (QW) diode lasers with GaInAsP/GaAs alloys operating at 980 nm were successfully produced in the "chimney" reactor.

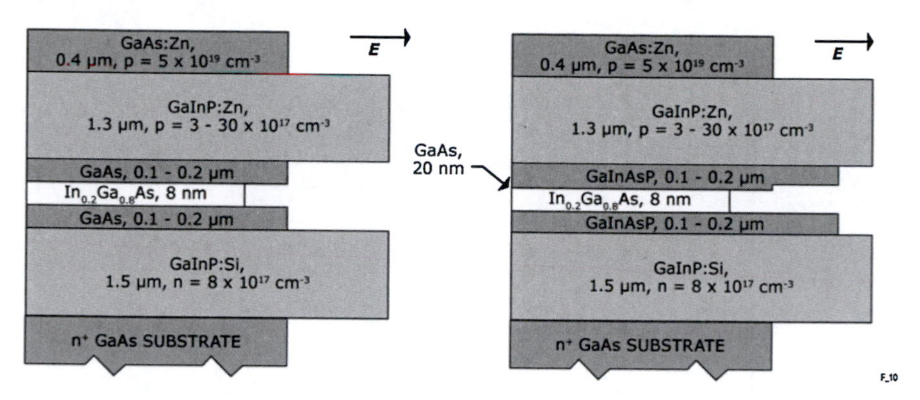

Fig. 2.10 Single quantum well separate confinement heterostructure device designs: on the left, GaAs optical waveguide material, and, on the right, GaInAsP (Energy gap = 1.7 eV) optical waveguide material. Profiles on the right edge show bandgap variation. Layer widths are not to scale [11]

Two types of separate confinement heterostructure QW devices were produced, differing in the optical-waveguide-layer material.

These structures were successfully mass transported and buried heterostructures devices were made. Ridge wave guide lasers with single spatial mode behavior to greater than 50 mW per facet output have also been fabricated from the structures with GaInAsP. These results demonstrated the ability of the "chimney" reactor to produce the state-of-the-art uniform and abrupt GaInAsP materials and laser diodes to near parity with those of AlGaAs alloys (Figs. 2.10 and 2.11).

Accomplishments and Insights

Produced for the first time abrupt interfaces in arsenides and phosphides grown by OMCVD. Modeled and designed a production OMCVD reactor. Produced first GaInP buried heterostructure lasers.

While working at MIT Lincoln Labs made me the engineer I am today, it did not afford me opportunities to develop or hone my social skills, something I identified as a critical gap as my engineering journey and leadership responsibilities progressed.

2.3.3 Gas Source Molecular Beam Epitaxy

Gas source MBE materials were developed at MIT Lincoln Laboratories in the 1990s with my colleague Dr. Paul A. Maki. The narrowest intrinsic photoluminescence linewidths 0.7 meV were obtained by solid source MBE-grown single quantum well structures. These linewidths were obtained by the elimination of impurities and controlling the stoichiometry at the quantum well interface. These impurities were significantly reduced by thermal conversion of the substrate in combination with controlling the V:III flux ratio when growing the quantum well. REF.

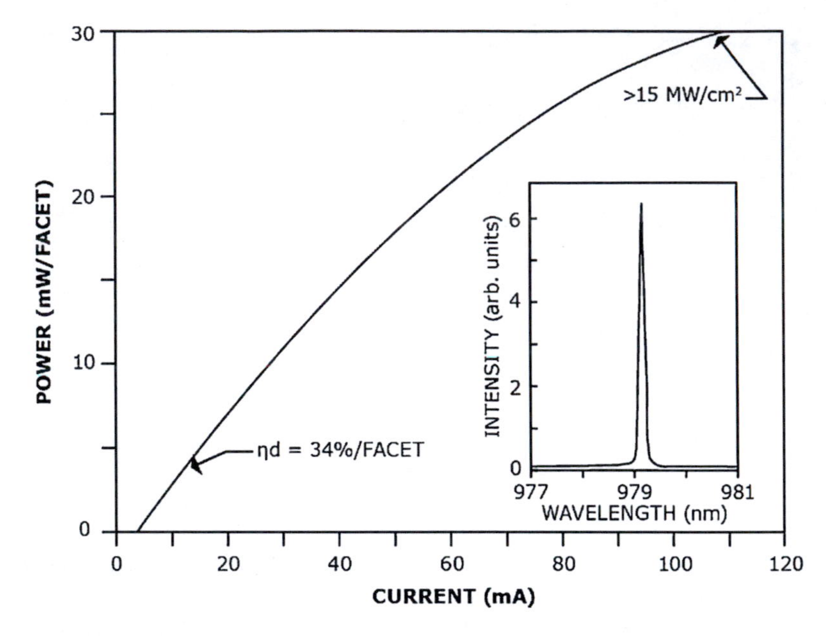

Fig. 2.11 Power per facet versus current in pulsed operation for buried heterostructure device, 1–2 μm wide and 350 μm long, with GaAs optical waveguide layers. The inset shows the emission spectrum at 10 mA with a single longitudinal mode near 980-nm [11]

Studies of growth kinetics between solid source MBE, OMCVD, and GSMBE shed insight into the crystal growth kinetics. The highest purity material was grown by solid source MBE, while the most abrupt interfaces were produced by GSMBE (Fig. 2.12).

An in-depth understanding of interface growth for solid source MBE, gas source MBE, and OMCVD was developed during this time given access to three epitaxial reactors. We were able to explore the material systems referenced below in Fig. 2.13.

Accomplishments and Insights

Characterized III–V material by three different growth techniques: MBE, OMCVD, and GSMBE. We developed an in-depth understanding of the bulk and interface growth and thermodynamics for III–V materials.

2.4 193-nm Lithography—Control and Quantification of Line-Edge Roughness in Resist Patterns

MIT Lincoln Laboratories had a center of excellence for 248, 193, and 157-nm lithography in the 1990s. SEMATECH funded a consortium that supported advanced development and characterization of lithography and dry etch development. A positive tone silylation process was extensively characterized with regards to exposure and defocus

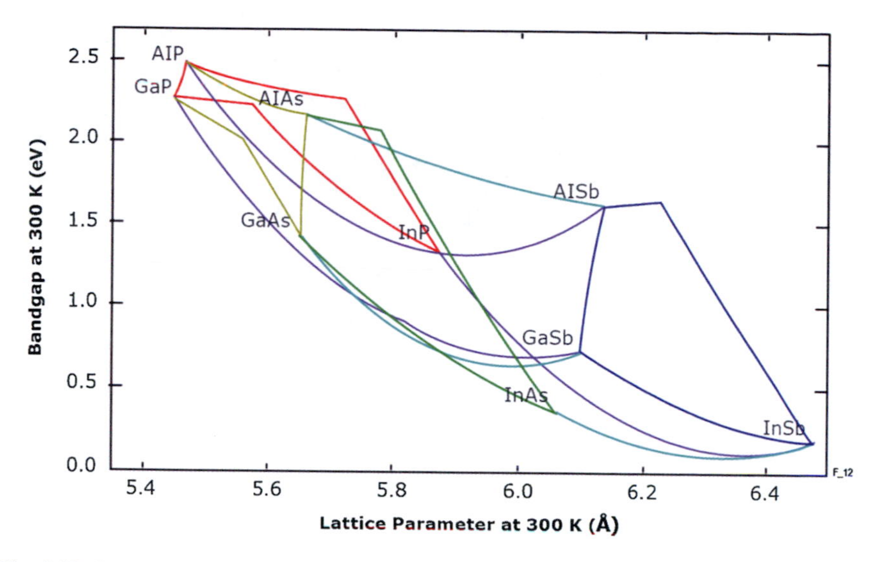

Fig. 2.12 Band gap versus lattice spacing at 300K for III–V semiconductors [12]

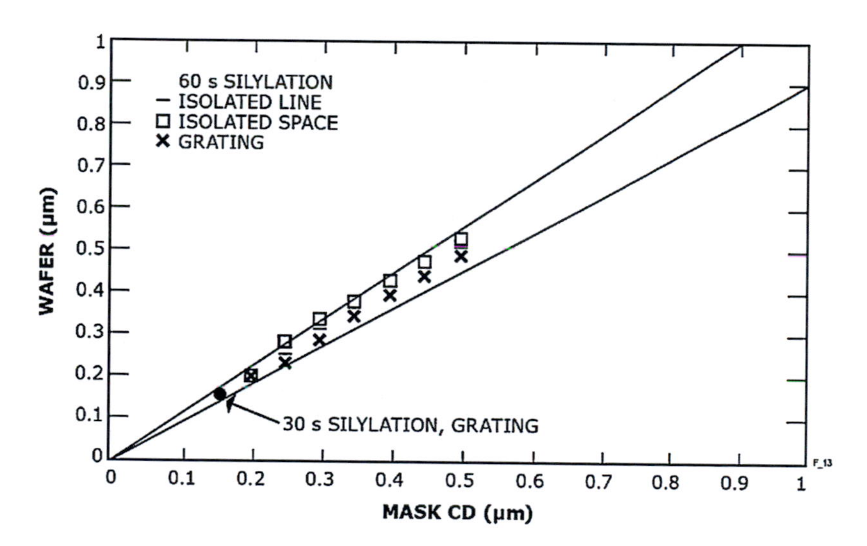

Fig. 2.13 Silylation process linearity at best dose (100 dose units –400 mJ/cm²). Resist O.76-j. tm-thick PVP silylated with DMSDMA at 90 °C and 25 Ton for 60 s was etched under optimized conditions with a 100% over etch. The grating linearity is extended to 0.175 im for a 30 s silylation time and 25% over etch. The two solid lines are the acceptable performance, i.e., deviations from the nominal feature size [13]

Gratings	K-Factor	Defocus (μm)	Dose Units						Exposure Latitude
			100	110	121	133	146	161	
250 nm	0.65	1.2 1.0 0.8 0.6 0.4 0.2 0 -0.2 -0.4	0.8	1.6	0.4	0.2			30%
200 nm	0.52	0.8 0.6 0.4 0.2 0 -0.2		0.2	1.0	1.0			20%
175 nm	0.42	0.6 0.4 0.2 0 -0.2 -0.4				0.2	1.0	0.6	20%

Fig. 2.14 Experimental exposure-dose matrix for 0.25, 0.20, and 0.175-μm gratings, (±10% CD). Each 193-nm dose unit corresponds to 1 mJ/cm^2. Resist 0.50 μm thick PVP silylated DMSDMA at 90 C and 25 Torr for 30 s was etched under optimized conditions with a 25% over etch [13]

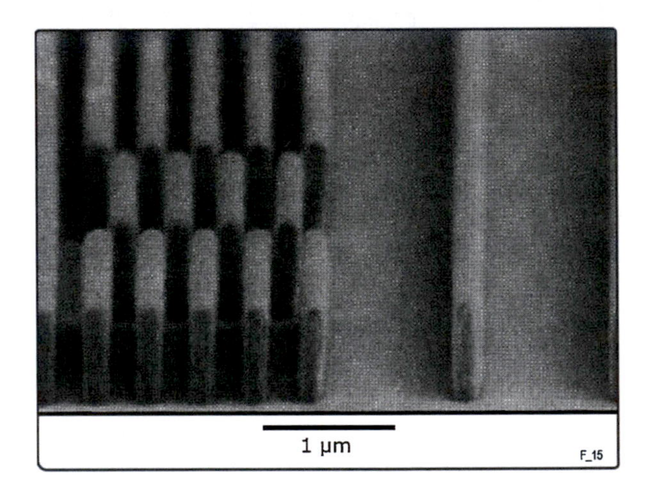

Fig. 2.15 Scanning electron micrograph of 0.02-μm gratings and isolated line [13]

latitudes, linearity of gratings down to 175 μm, and resolution of 0.15-μm gratings and isolated lines. These results are summarized in Figs. 2.13, 2.14, and 2.15.

We characterized line-edge roughness in single layer, top-surface imaging, bilayer, and trilayer resist schemes. The results indicated that in dry-developed resists there is inherent line-edge roughness that results from the etch mask, resist erosion, and their dependence on plasma etch conditions. In top surface imaging, the abruptness of the etch mask, i.e., the silylation contrast, and the silicon content in the silylated areas are the most significant contributors to line-edge roughness. The mechanism and magnitude of the line-edge roughness are different for different resist schemes, and require specific optimization. Plasma etching of silicon, like oxygen dry development, contributes to the final line-edge roughness of patterned features (Figs. 2.16 and 2.17).

Process	Resist	Exposure Tool	Development	LER Result (Top Down SEM)		
				σ (nm)	Indicated Range (nm)	Period (nm)
SLR	UV5	248	Wet	2	9	>100
SLR	IBM V1.1	193	Wet	4	17	>200
TSI	MX-P7	193	Dry (O_2)	5	25	<90
TSI	MX-P7	193	Dry (C_2F_6+O_2)	2	9	>100
TSI	NEK 304	193	Dry (CF_4+O_2)	4	15	>100
Bilayer	Experimental	248	Dry (O_2)	2	6	>200
Bilayer	Experimental	193	Dry (O_2)	2	9	>200
Trilayer[a]	UV5	248	Dry (O_2)	1	8	>100

F_16

Fig. 2.16 Table of line-edge roughness measured for various resist schemes. Measurements taken on a 175-nm isolated line (193-nm) and 250-300-nm isolated line (248-nm) at best dose and focus except for (a), which was taken on a 100-nm isolated line. All wafers shot with Cr mask except for (a) which is Cr-less phase shift [14]

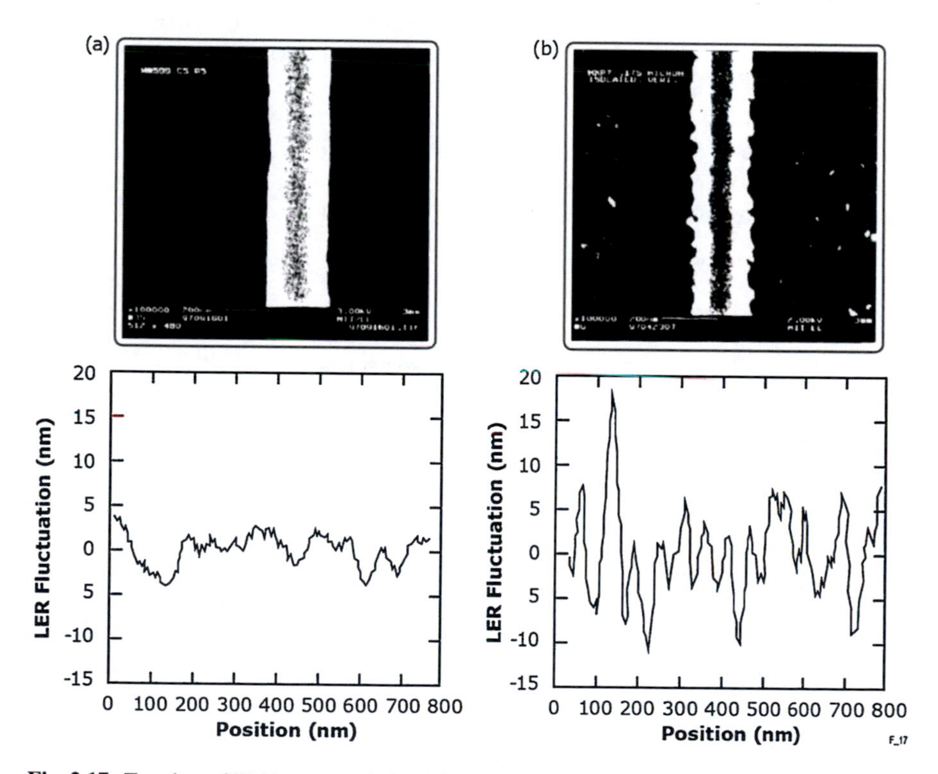

Fig. 2.17 Top-down SEM images and plots of LER fluctuations for nominal 175-nm isolated lines in (**a**) TSI (MX-P7) with a C_2F_6 + O_2 breakthrough etch and (**b**) TSI (MX-P7) without a breakthrough etch. These images and LER plots correspond to the resist schemes listed in Fig. 2.16 [14]

SEM and AFM techniques to measure and analyze edge roughness in patterned resist and silicon features was developed. Both techniques are able to follow small changes in line-edge roughness. The measurement repeatability of SEM and AFM was characterized and are 0.1 and 0.6-nm, respectively. These advancements in photolithography and characterization of line-edge roughness enabled the industry to model and evaluate the effects of these parameters on transistor performance.

Accomplishments and Insights

I made significant contributions to 257, 193 and 157-nm lithography and dry etch—in particular, measurement capabilities. Mentorship and significant funding from Karen Brown from SEMATECH was instrumental in achieving these contributions. I developed a deep understanding of design of experiments and statistical process control during this period of my career.

2.5 980 and 1480-nm Laser Diode Manufacturing

I joined Corning Lasertron in 1999 to ramp-to-rate 980 and 1480-nm laser diodes into manufacturing. Initially, I was hired to evaluate the ability of the existing OMCVD reactor to produce 1480-nm pump lasers. I redesigned this reactor to produce arsenides and phosphides for laser diodes. In a short time, I became the Director of Materials for both 980 and 1480-nm laser diodes. This marked my transition into managing and leading a group. We were evaluating material supply from three sources—internal at Lasertron facility, internal at Corning facility and external IQE. This position honed the skills of material and device correlations as shown in Figs. 2.18 and 2.19. Instituting

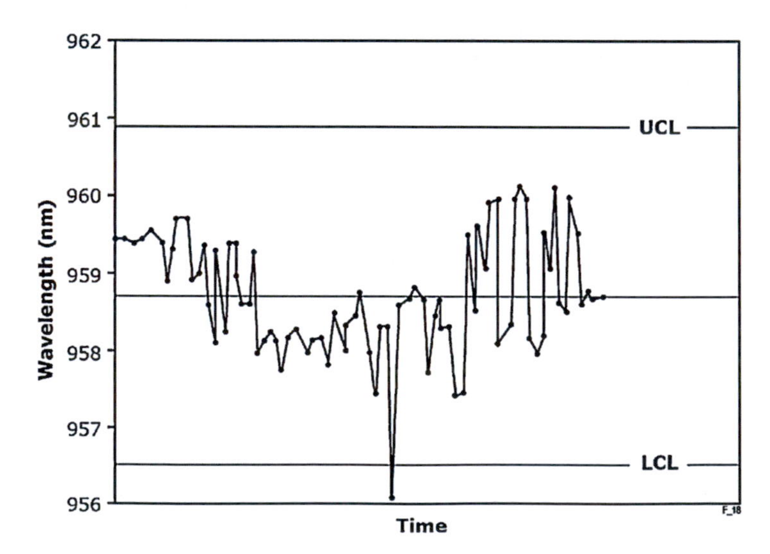

Fig. 2.18 Uncorrected wavelength data for standard sample over an eight-month period. Control window 6-nm [15]

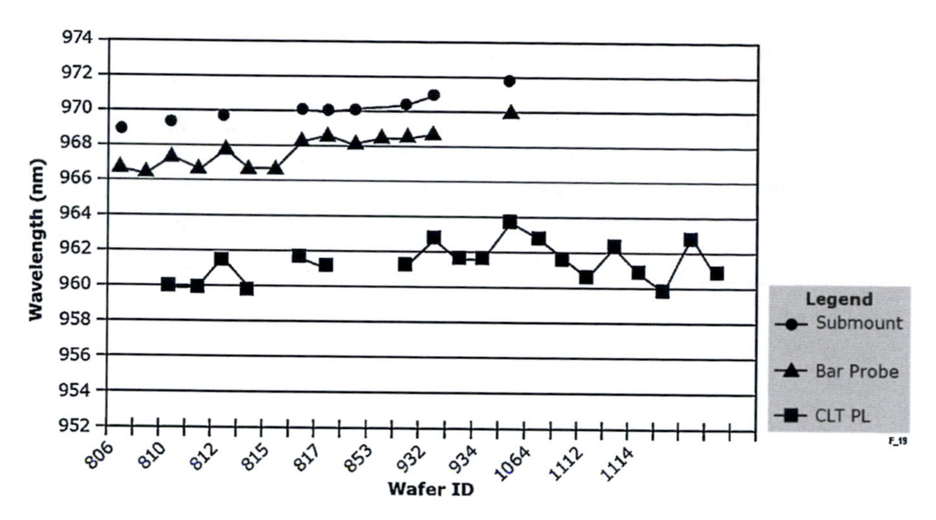

Fig. 2.19 Wavelength correlation between PL, chip, and sub mount measurement. A linear regression analysis shows $R^2 = 0.84$ between PL and chip, and $R^2 = 0.89$ between PL and sub mount [15]

control limits and understanding the correlations between material parameters and device parameters was critical to producing these laser diodes in volume.

Accomplishments and Insights
It was during this time at Corning Lasertron that I transitioned into a management role. I enjoyed being technical, creating, and leading teams. I did not look back to being an individual contributor.

2.6 The EPI Pack—Boy This Was Fun

In 2003 after the Corning Lasertron facility was shut down and technology transferred to Alcatel in France, I started my own consulting company: The EPI pack. The goal was to provide a service to the semiconductor industry to specify and qualify epitaxial material. I already had one client: Axsun Technologies. Although the work was challenging, I really was enjoying working for myself when I got a call from a colleague asking me to interview at BAE Systems. He said he had the perfect job for me: ramp-to-rate for night vision sensor technology into thermal weapon sight systems.

Accomplishments and Insights
I could run my own business, but choose to go back to the corporate side. I was used to a large structured organization.

2.7 Night Vision Systems

I joined the Sensor and Soldier Solutions sector of BAE Systems in 2004, managing the foundry ramp-to-rate for night vision sensors. We ramped vanadium oxide microbolometers from a few wafers per month to 300 wafers per month. In 5 years, we delivered over 100,000 night vision sensor systems. We transitioned from a pixel size of 46 microns to 28 microns to 17 microns and finally 12 microns, pushing the size, weight, power, and cost performance to the limit. During this time (2004–2015), I held three different positions: Foundry Manager, Product Line Director, and Program Director at three different locations—Lexington, Massachusetts, Manassas, Virginia, and Hudson, New Hampshire (Figs. 2.20 and 2.21).

Accomplishments and Insights

In this position, I transitioned into a leadership role, created a foundry organization that could produce low-cost, high-volume uncooled sensors for night vision systems, and lead the organization from making one to over 5000 units per month.

My boss said one day, "Susan you have it all," but you need more business sense. He sent me to the Tucks School at Dartmouth College. This gave me the ability to translate between the technical people and the management side.

In 2015, I became the Foundries Business Area Engineering Director for the Electronic Systems Sector for GaAs and Gallium Nitride (GaN) semiconductor products, including night vision sensors. This has been the most challenging, but rewarding transformation project of my career. It came full circle. I started out my career with this group of experts in microwave design and products back in 1983 at

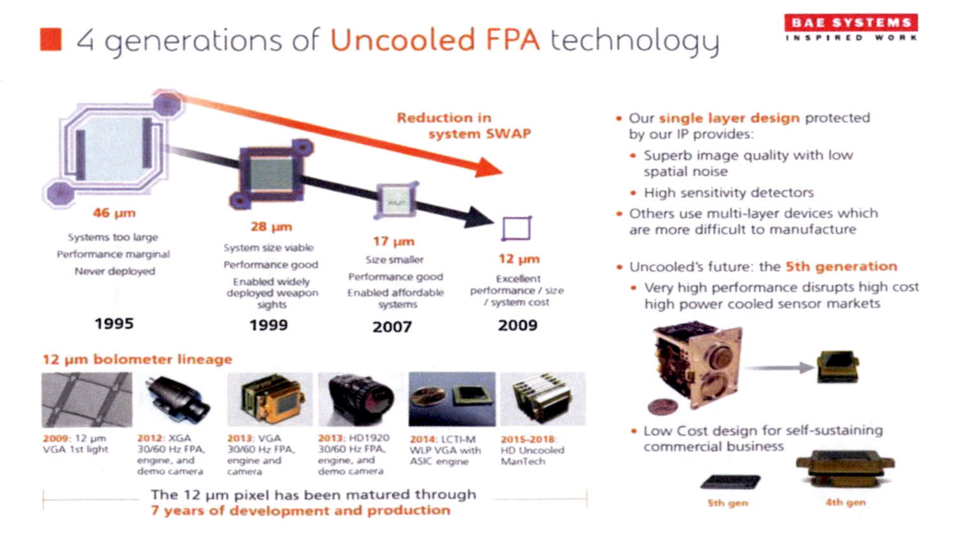

Fig. 2.20 We pushed the performance of night vision systems—size weight and power by going to smaller pixel sizes (BAE Systems)

Fig. 2.21 Night vision system head mounted device. *Night vision goggles. Image by SAC Chris Hill/MOD* Open Government Licence v1.0, via Wikimedia Commons

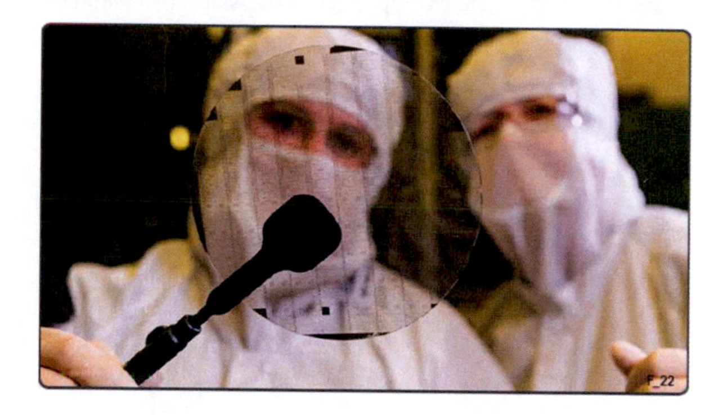

Fig. 2.22 Process engineers viewing a 6 in. GaN wafer. https://www.baesystems.com/en-us/product/foundry-services

General Electric. Back then, they were brilliant experts and 35 years later, they were still brilliant, but not disciplined in producing products reliably and consistently. My job was to transform this organization. At the writing of this chapter, we are 80% of the way there. We went from 35 foundry-focused people to 260 people creating Monolithic Microwave Integrated Circuits (MMIC) products in 3 years.

Training the people to understand Six Sigma, process control, yield improvement, and new product introduction was very exciting, but very challenging—especially when the ramp-to-rate was right upon us. We positioned the organization to deliver in a high-mix, small-quantity environment. We deliver 250,000 die per year, including high-power and low-noise devices for mostly internal consumption for electronic warfare systems. We also offer foundry services for other U.S. Department of Defense customers, and we are developing next-generation technologies, in particular 0.18 μm GaN high-power amplifiers (Figs. 2.22, 2.23, and 2.24).

Parameter	0.1 μm PHEMT	70 nm PHEMT	50 nm MHEMT	0.18 μm NFP* GaN	0.18 μm NFP SC* GaN
FT	90 GHz	150 GHz	250 GHz	57 GHz	67 GHz
Fmax	225 GHz	310 GHz	1 THz	160 GHz	200 GHz
Gm	570 mS/mm	710 mS/mm	2000 mS/mm	370 mS/mm	435 mS/mm
Imax(Vgs)	615 mA/ mm(1V)	700 mA/ mm(0.8V)	800 mA/ mm(0.5V)	1200 mA/ mm(1V)	1450 mA/ mm(1V)
Vds max oper.	5V	4V	2V	30V	30V
BVgd	11V	10V	6V	>100V	100V
NFmin	1.4 dB @ 40 GHz	1.1 dB @ 40 GHz	0.8 dB @ 40 GHz	1.7 dB @ 40 GHz	1.6 dB @ 40 GHz
Wafer thickness	55 and 100μm	55 and 100 μm	55 and 100 μm	55 and 100 μm	55 and 100 μm
Backside vias (slot width)	55μm: 15μm 100μm: 30μm	55μm: 15μm 100μm: 30μm	55μm: 15μm 100μm: 30μm	55μm: 15μm 100μm: 30μm	55μm: 15μm 100μm: 60μm
Wafer size	6-inch	6-inch	3-inch	4-inch (55 μm)** 6-inch (100 μm)	4-inch (55 μm)** 6-inch (100 μm)

* FP – Field plate, NFP – No field plate, SC – Scaled channel
** 6-inch will be available in 2019, these two processes can be run concurrently

Fig. 2.23 Monolithic microwave integrated circuits processes and performance offered as foundry service by BAE Systems. https://www.baesystems.com/en-us/product/foundry-services

Fig. 2.24 Our MMICs provide the heart of the electronic warfare systems produced by BAE Systems to protect our warfighters and our country (BAE Systems)

Accomplishments and Insights

This is the most challenging project of my career, transforming a research and design organization into a product development delivery-focused organization.

Acknowledgements To my parents, who provided me the freedom to do anything I wanted.

I want to thank all the individuals who mentored me and pushed me, and those that made it challenging for a female in engineering—for me, it made me stronger. Secondly, I would like to acknowledge the great organizations that I have worked for and the resources and support that they provided.

I am eternally grateful for a second chance at life following a tragic car accident in 1991 in which I beat the odds, but with a collapsed lung, broken neck, and paralysis. The near-death experience was a gift that served to make me understand what is important in life. After the accident, I had my two daughters—and I treasure each day I have in this physical world.

Final Insights

A positive attitude brings much fortune in life—it is all a state of mind.
The most important thing is a partner who treats you as an equal.

What Has Influenced Me?

Gloria Steinham's response to a young woman struggling with career and raising children, who asked her, "How do we do it all?" at a New Hampshire Women's conference. Gloria's response: "At least you have a choice."

We all have choices—make them and learn from them.

References[1]

1. Palmateer SC (1982) Manganese redistribution in molecular beam epitaxial grown GaAs. Master Thesis, Electrical Engineering, Cornell University
2. Palmateer SC (1985) Mechanisms of impurity redistribution in gallium arsenide substrates and molecular beam epitaxially grown layers. PhD Thesis, Electrophysics, Cornell University
3. Palmateer SC, Eastman LF, Calawa AR (1984) The use of substrate annealing as a gettering technique prior to MBE growth. J Vac Sci Technol B 2:188
4. Palmateer SC, Maki PA, Katz W, Calawa AR, Hwang JCM, Eastman LF (1984) The influence of V:III flux ratio on unintentional impurity incorporation during molecular beam epitaxial growth. In: Proceedings of the 1984 international symposium on gallium arsenide and related compounds, Biarritz, France, 26–28 Sept 1984
5. Palmateer SC, Maki P, Hollis M, Eastman LF, Ward I, Hitzman C (1982) Growth of planar doped barrier structures in gallium arsenide by molecular beam epitaxy. In: Proceedings of the 1982 international symposium on gallium arsenide and related compounds, Albuquerque, NM, Sept 1982
6. Chao PC, Palmateer SC, Smith PM, Mishra UK, Duh KHG, Hwang JCM (1985) Millimeter-wave low-noise high electron mobility transistors. IEEE Electron Device Lett EDL-6(10):531–533
7. Wang CA, Groves SH, Palmateer SC, Weyburne DW, Brown RA (1986) Flow visualization studies for the optimization of OMVPE reactor design. J Cryst Growth 77:136–143
8. Wang CA (1990) The Lincoln laboratory. J Spring 3(1)
9. Palmateer SC, Groves SH, Wang CA, Weyburne DW, Brown RA (1986) Use of flow visualization and tracer gas studies for designing an InP/InGaAsP OMVPE reactor. In: Proceeding of the 3rd NATO workshop on materials aspects of InP, Harwichport, MA, 22–25 Sept 1986
10. Palmateer SC, Groves SH, Caunt JW, Hovey DL (1989) New OMVPE reactor for large area uniform deposition of InP and related alloys. J Electron Mater 18:645

[1] Dr. Palmateer has over 140 Publications and Presentations.

11. Liau ZL, Palmateer SC, Groves SH, Walpole JN, Missagia LJ (1992) Low-threshold InGaAs strained-layer quantum-well lasers (l=0.98 mm) with GaInP cladding layers and mass-transported buried heterostructure. Appl Phys Lett 60:6
12. (2013) Plot bandgap versus lattice constant. Open band diagrams software. Scott J Maddox. http://scott-maddox.githun.io/openbandparams/0.7/examples/_Plot_Bandgap_vs_Lattice
13. Palmateer SC, Kunz RR, Horn MW, Forte AR, Rothschild M (1995) Optimization of a 193-nm silylation process for sub-0.25-um lithography. Proc SPIE 2438:455
14. Palmateer SC, Forte AR, Kunz RR, Horn MW (1996) Dry development of sub-0.25 um features patterned with 193 nm silylation resist. J Vac Sci Technol A 14:1132
15. Nelson CM, Palmateer SC, Forte AR, Lyszczarz TM (1999) Comparison of metrology methods for quantifying the line edge roughness of patterned features. J Vac Sci Technol B B 17(6):2488–2498

Chapter 3
Thin-Film Deposition of Hybrid Materials

Adrienne D. Stiff-Roberts

3.1 Motivation for Hybrid Thin Films in Electronic/Optoelectronic Devices

Some of the most exciting materials research in the twenty-first century attempts to resolve the challenge of designing new materials with unique properties from first principles. Such custom materials are critical to enabling systems of the future, like wearable electronics, flexible and transparent displays, or devices for solar energy conversion and storage. These applications require materials with more versatility, more integrated functions, and more environmentally responsible processing, especially compared to traditional options (i.e., inorganic semiconductors). My goal as a researcher is to help lead the transition from traditional to emerging materials by investigating a novel thin-film deposition technique to control structure and properties in heterogeneous systems for electronic devices (based on the control of electrons) or optoelectronic devices (based on the interaction of light and matter).

To be sure, traditional semiconductors, like silicon or gallium arsenide (GaAs), are ubiquitous in electronic and optoelectronic systems for a reason. These crystalline, inorganic materials have beneficial properties, such as high-speed response, efficient conversion of light energy, and stability over time. Even though inorganic semiconductors can be rare, and the ultra-high vacuum and high-temperature conditions required for material processing can be expensive, these materials dominate industrial and commercial products because the device areas required for typical applications are very small. Yet, we can imagine many electronic or optoelectronic systems that would benefit from larger device areas. For example, biomedical sensors that detect changes in the optical properties of biological samples can be used to diagnose diseases. High-resolution digital displays that emit bright light with

A. D. Stiff-Roberts (✉)
Jeffrey N. Vinik Professor, Department of Electrical and Computer Engineering, Duke University, Durham, NC, USA
e-mail: adrienne.stiffroberts@duke.edu

A. C. Parker, L. Lunardi (eds.), *Women in Microelectronics*, Women in Engineering and Science, https://doi.org/10.1007/978-3-030-46377-9_3

excellent color quality are important components of consumer electronics. Solar cells that convert absorbed sunlight into electricity are critical for meeting the energy needs of the future. For each of these applications, inexpensive, large-area devices with flexible form factors could be game changers. In addition, the ability to combine multiple devices with very different functions on a single substrate, like paper, textiles, or glass, could enable new integrated systems. Organic semiconductors, which comprise synthetic, carbon-based molecules, are well suited to these desired, forward-looking characteristics. Although usually thought of as insulators, certain small molecules and polymers can demonstrate the definitive characteristic of a semiconductor, namely, an energetic bandgap across which light can be absorbed and emitted and which enables the conduction of charge as a function of temperature. Organic materials provide efficient conversion of light energy, enable designer functionality and optical properties, and offer the potential for synthesis using sustainable and non-toxic chemistry with inexpensive, earth-abundant raw materials. However, organic components are not ideal for high-speed device requirements, and they are less stable in the environment (especially ultraviolet light, moisture, and oxygen).

In contrast, *hybrid* materials that comprise organic and inorganic constituents can combine disparate properties to mitigate the trade-offs that exist for any single material type, thereby creating new heterogeneous materials with tailored characteristics. For example, hybrid organic-inorganic materials can impart multi-functionality, flexibility, transparency, and stability to devices. A critically important requirement to realize the promise of hybrid materials for devices is to understand and control thin-film deposition, which faces several obstacles. For example, the electronic and photonic properties of organic components often change from the as-synthesized characteristics when the material is incorporated into a film. In addition, the ability to predict material properties and device performance is often hampered by variations in film behavior that depend on the deposition technique. This inconsistency complicates the comparison of results from different labs that is required to make progress for a given application. Furthermore, while many hybrid materials benefit from simple solution processing that is amenable to large-area, roll-to-roll manufacturing, it has proven difficult to control material uniformity such that high-yield, high-throughput production is achieved. Fundamentally, the primary technological challenge facing thin-film deposition of hybrid materials is to control the co-deposition of two or more materials with different natures such that a new material is synthesized with pre-determined properties and functionality.

My work to address this challenge resides at the nexus of materials science, physics, and electrical engineering, and it has roots in my doctoral research on the thin-film growth of inorganic semiconductor nanomaterials, which later expanded to include organic and hybrid materials as a junior faculty member at Duke University. While the logic of my research progression is clear in hindsight, my journey to this point was not driven by a technology roadmap or an engineering grand challenge. Instead, when I reflect on the interests I have pursued and the choices I have made, three guiding principles are clearly responsible for my current circumstance: (1) a passion for learning, (2) an imperative to contribute new knowledge, and (3) a responsibility to help others.

3.2 Lifelong Love of Learning

I knew early on that I wanted to be a professor. I did not know which discipline I would study, and I had no idea what research was, but I knew I wanted to teach at the highest level possible. Earlier dreams of becoming an astronaut or a heart surgeon were dashed by the harsh realities of the Space Shuttle Challenger disaster and the dissection of a pig's heart in the fifth grade. I started to look for a career that would be important and could help people, but with less risk and less gore. When I thought about what I enjoyed most, the answer was school. I enjoyed learning how the world around me worked, and because I could not be a student forever, the next best thing was to be a teacher.

Like many people outside of academia, initially I thought the extent of being a professor was to teach a few classes during the academic year, take summer vacations, and earn a higher salary than a K-12 teacher earns. Unlike many people outside of academia, I had exposure to university settings and university life to help correct this false impression. My father, Lee V. Stiff, was a mathematics education professor at North Carolina State University (NCSU), and my mother, Renee Flood Stiff, was a university librarian at North Carolina Central University (NCCU). I accompanied them to work from time to time, and I experienced how their jobs enabled our comfortable home life. The most impactful impressions of what it meant to be a professor came from my father when he explained his job to me. He told me that professors get paid to sit around and think and that professors are their own boss and can decide how to spend their time at work. This insight, combined with my desire to teach, convinced me that I could sit and think for a living, and by the time I was in high school, I had set my sights on becoming a professor.

3.2.1 K-12 Education

Having the knowledge that I wanted to become a professor pre-determined many of my activities in school. To be a professor, I knew I needed a PhD, which meant I had to go to college, which meant I had to excel in high school, which meant I had to always do my best at any endeavor. My parents were very consistent in sharing that message, and it allowed me to work hard and challenge myself—not in a stressful way through competition with or comparison to others, but in a constructive way through meeting high expectations that I set for myself. As a result, I was strong in all of my subjects at school, and there was no obvious area in which I should pursue my graduate education. Science was not my specialty or core interest. I was active in student government, I played varsity basketball, and I took lessons in piano and tap dancing. Throughout middle school and the beginning of high school, I enjoyed science as much as any other subject or activity. In fact, my father always thought that I would become a lawyer.

In looking back, there were probably two main influences that lead to my selection of physics as a major in college. The first influence was obvious. During my junior year, when it became necessary to select a major in order to determine which schools and scholarships I would apply to, my favorite class was an advanced placement physics course. This was due in large part to the teacher of the course, Ms. Elizabeth Woolard, who took great pains to provide in-class demonstrations and example problems that made it clear how relevant physics is to our everyday lives. I was always amazed when common experiences could be explained by the physics principles we were learning in class, and I felt that I could not go wrong by concentrating on physics as a fundamental discipline upon which other fields of science are built. Moreover, it did not hurt to have gentle nudges from my parents who recognized that pursuing science would be a beneficial direction for my future. Interestingly, I did not really appreciate the second main influence until much later in life—the reason that I was not intimidated by science, especially as a black woman, and the reason that I considered science careers as my dream jobs, even in childhood. I was always a fan of science fiction, especially stories about space. I watched *Dr. Who* and *Star Trek: The Next Generation*. Some of my favorite books were *The Tripods Trilogy* by John Christopher, *The Illustrated Man* by Ray Bradbury, and *Parable of the Sower* by Octavia Butler. My love of science fiction and interest in the ideas of space travel and extraterrestrial life had indoctrinated me to consider science as an essential tool and skill for the future. I thought that learning science was fun, so the idea of becoming a physics professor seemed completely natural to me.

Engineering, however, was more of a mystery. I did not even know what engineering was until my senior year in high school. I had seen pamphlets about engineering at college fairs, and I knew it had something to do with science, but I wanted to understand more. I joined the Pre-College Initiative group of the National Society of Black Engineers (NSBE) that was active at my high school and began to appreciate that engineering is the application of science to solve real problems. The school advisor for the NSBE chapter, Mr. Tom Blanford, also taught me in an elective course on lasers. I am sure that we learned something about how lasers work, but the part of the course that stuck with me was using mirrors to get the laser to follow a defined path around the classroom. It seems extremely simple now, but that was one of the coolest things I had ever done in school. The exposure to using lasers in that course planted a seed that the field of light interactions with matter was a natural link between the physics I loved and the practical applications pursued in engineering.

3.2.2 Undergraduate Education

One of the first, most important independent decisions I ever made was where to attend college. I had very different ideas about the experience I wanted in college compared to what my parents wanted for me. My parents attended college at predominantly white institutions (PWIs) when integration was still relatively new. They had attended segregated primary and secondary schools, and they had a mission to prove that they could compete and excel in any arena. In contrast, I had attended integrated public schools my whole life and had already demonstrated that I could perform at the highest level compared to my peers. I also dealt with the micro-aggressions of being the only black person in advanced and honors classes on a regular basis (although I did not have a name for it at that time). The one thing I knew was that I wanted to attend a Historically Black College/University (HBCU). I wanted to be in an environment where, whatever obstacles or challenges I faced, I would not have to wonder if my race had anything to do with it. Although my parents knew that HBCUs have a long tradition of preparing black students to excel, in some ways, they still viewed PWIs as the ultimate proving ground for convincing others that you could perform at the highest levels. In addition, while my parents implored me to understand that the real world would not be like an HBCU, I held fast to my determination to attend one.

I applied to several HBCUs with strong science and engineering programs, like Howard University, North Carolina A&T State University, and Florida A&M University. In the end, I chose to attend Spelman College, an HBCU for women in Atlanta, GA, which participated in the Atlanta University Center Dual-Degree Engineering Program (AUC DDEP). Through this program, I was able to pursue two undergraduate degrees in physics and electrical engineering from Spelman College and the Georgia Institute of Technology, respectively. I also had the opportunity to experience two very different institutions. One, a liberal arts college for black women in which I supplemented my physics studies with a broad education in humanities and social sciences, and one, a PWI engineering institution in which I had an intense focus on electrical engineering. In addition, choosing Spelman College had the added benefit that I was able to attend college in the same city as my high school sweetheart and future husband, Francis Roberts, Jr., who was a junior at Morehouse College when I began my studies at Spelman.

An important aspect of my undergraduate education was the full scholarship I received to attend Spelman College, the NASA Women in Science and Engineering (WISE) Scholarship. As a WISE scholar, I not only received financial support for my education, but I received invaluable mentorship from the faculty at Spelman College and community from the exceptional women in the program. Beginning with a summer preparatory program before my first year, as a WISE scholar, I was groomed to succeed at Spelman, at Georgia Tech in the AUC DDEP, and in graduate school. Led by the esteemed Prof. Etta Falconer, the WISE program challenged me academically, equipped me with the skills to survive and thrive in different environments, and encouraged me to pursue a graduate degree in science or engineering. The WISE scholarship included a research internship at NASA over the summer. As a result, I spent two summers at the NASA Ames Research Center in Mountain View, CA (Fig. 3.1). During my first summer, I performed data processing for a psychophysiology lab that sought to teach astronauts to control their physiological responses to stress in order to reduce sickness in a weightless environment. During my second summer, I designed and tested a system for measuring the amount of food eaten by mice under experiment in space using an accelerometer circuit to

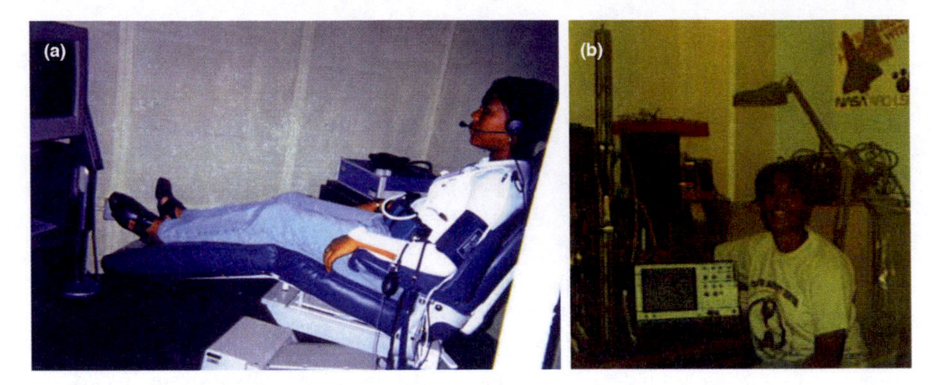

Fig. 3.1 (**a**) Photograph of me wearing sensors used to monitor physiological responses in the autogenic feedback training experiment of the psycho-physiological lab at NASA Ames Research Center in 1995; and (**b**) photograph of me with the food bar plate mass system I designed at NASA Ames Research Center in 1996

determine the natural frequency of a food bar plate. These experiences were my first introductions to research as a concept. In fact, it was not until my second summer at NASA that I started to understand that research is a requirement of being a professor, and in many cases, the primary responsibility.

Overall, the AUC DDEP was ideal in that it allowed me to attend an HBCU in the Black Ivy League and a top institution in engineering (both of which helped alleviate my parents' concerns), and I was able to excel in both. That is not to say that I did not experience any hardships or challenges. I had to learn to adapt to different environments, different styles of teaching, even different academic calendars, with Spelman on a semester system and Georgia Tech on a quarter system. One of my biggest challenges was the quantum mechanics course I took at Morehouse College. Taught from a purely mathematical perspective, and covering material that is non-intuitive, I struggled to make the connections with everyday experiences that helped make sense of complex ideas and that drew me to physics in the first place. While I survived the course, quantum mechanics remained a weakness upon my entrance to graduate school. Nonetheless, my experience at Spelman College vindicated my choice to attend an HBCU. The confidence I developed by being able to focus on my education without the trappings of racial microaggressions was engrained by the time I left Spelman, and it sustained me even when the hiatus was over. My undergraduate experience also solidified my path in multi-disciplinary pursuits. The AUC DDEP helped me appreciate the value of combining disciplines to gain a better understanding of fundamental topics, and my study of physics and electrical engineering, with a focus in optics (inspired by my high school course on lasers), prepared me to delve deeper into light interactions with matter.

3.2.3 Graduate Education

Once I understood that a core component of being a professor is to conduct research, I knew I was most interested in research that would help solve problems or have application to everyday life. Most of the physics research I had encountered was too esoteric, so I chose to pursue my PhD in Applied Physics at the University of Michigan, Ann Arbor. One important reason I chose to join this program is that the director, Prof. Roy Clarke, had successfully created a diverse and inclusive environment in which I could find community. A second important reason was that, beyond the broad topic of optics, I had no idea in which area of research I was most interested. I spent the summer before beginning graduate school as a Research Fellow at AT&T Labs - Research in Red Bank, NJ, and in addition to my own research on optical fiber communication, I learned about research from other students in the program, such as microelectromechanical systems and personal digital assistants. While I gained a lifelong mentor from that experience, Prof. Leda Lunardi, I still did not find a research passion. In contrast to many people I met, those driven to solve a specific research problem or to become an expert in a specific research discipline, I viewed the PhD as a means to an end, and research was a required step I had to complete. The Applied Physics program at the University of Michigan afforded me the flexibility to study core physics concepts at the graduate level yet join a research group in any discipline across campus that was related to physics. Given the enormous size of the University of Michigan, I knew I would be able to find *something* to which I could dedicate 5–6 years of my life.

The first year of graduate school was a time of significant change for me. I moved to the mid-west, lived without a roommate, and had to deal with snow on a regular basis, all for the first time. Moreover, studying for my courses required an intensity and focus matched by few other times in my life. My old friend, quantum mechanics, was especially challenging. I spent hours upon hours reading course material, practicing problems with others, attending office hours, reviewing notes, and completing homework sets. Eventually, all the hard work began to pay dividends as the abstract ideas finally began to make sense. Around the same time, I attended a weekly seminar in which faculty affiliated with the Applied Physics program presented their work to help recruit students to join their groups. There had been some interesting talks related to optics, but nothing that really captured my imagination. Then, I heard a talk by Prof. Pallab Bhattacharya on quantum dot lasers. During his talk, I was fascinated by the fact that he could make a three-dimensional version of the particle-in-a-box problem I had studied in quantum mechanics. These so-called quantum dots are nanomaterials that exhibit quantum confinement in three dimensions. Furthermore, he used the quantum dots to make lasers for fiber optic communication, topics that piqued my interest since high school and were further developed at Georgia Tech and AT&T Labs—Research. I was hooked, finally finding a research area that, in my view, perfectly blended physics and practical applications, and I joined his group within a few weeks of the talk. As a member of the Bhattacharya group, I learned to make these inorganic quantum dots by an ultra-high-vacuum crystal growth process

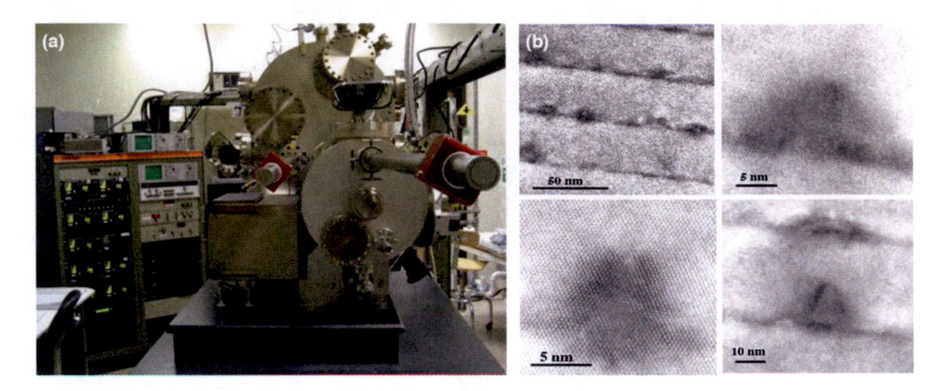

Fig. 3.2 (**a**) Photograph of MBE system from Prof. Pallab Bhattacharya's group at the University of Michigan, Ann Arbor circa 2000; and (**b**) cross-sectional transmission electron microscopy images of InAs/GaAs quantum dots I grew using MBE

known as molecular beam epitaxy (MBE), thus beginning my introduction to materials science and engineering. Epitaxial growth of III–V compound semiconductor nanomaterials, namely, InAs/GaAs quantum dots (Fig. 3.2), opened the door to physical vapor deposition, materials characterization, and the materials science concept of structure-property relationships.

After my second year of graduate school, I got married, and my husband moved to Ann Arbor, MI. Managing the new realities of married life, while maintaining the same intensity in my research, was a challenge. I am fortunate that my life partner has always been supportive of my career choice and understanding of the complexity it brings. As I progressed through my doctoral research, I focused on quantum dots grown by MBE as light absorbers for infrared (IR) photodetection [1]. This research was important to enable sophisticated military imaging systems that operate near room temperature with the maximum spatial and spectral resolutions possible.

Quantum dots enable multi-spectral IR photodetection by controlling light absorption through the dot size, shape, or material composition. Quantum dots also significantly decrease thermally generated dark current (or noise) in IR photodetectors due to their quasi-zero-dimensional size, which increases the signal-to-noise ratio for devices, as well as the resulting operating temperature. Near-room-temperature operation of IR photodetectors would eliminate the need to cool imaging systems with liquid nitrogen in the field, which has significant impact on the cost, weight, and deployability of such systems. During my PhD, I demonstrated IR imaging in quantum dot IR photodetector arrays for the first time [2], described the device physics of quantum dot IR photodetectors [3], and established the materials and device requirements for state-of-the-art performance [4]. Thus, completing my PhD in 2004 (Fig. 3.3) set the stage for my broader independent research emphasis on thin-film deposition of semiconductor materials for optoelectronic devices.

Fig. 3.3 Photograph of my mother, father, and me during my hooding ceremony after completion of my PhD in Applied Physics at the University of Michigan, Ann Arbor in 2004

3.3 Contribution to New Knowledge: Resonant Infrared Matrix-Assisted Pulsed Laser Evaporation of Hybrid Thin Films

I joined Duke University as an Assistant Professor in the Department of Electrical and Computer Engineering in 2004. At that time, it was not uncommon for fresh PhDs in engineering to go directly to a faculty position. Even though (or perhaps because) I was a North Carolina native, I never envisioned myself as a faculty member at Duke. However, during my job search, the recruitment efforts of the department chair, Prof. April Brown, and dean, Prof. Kristina Johnson, convinced me to try it. They created a welcoming and supportive environment and they worked diligently to solve the so-called "two-body problem" by helping to find opportunities for my husband within the university. I had finally achieved my long-held goal of becoming a professor, and one of the first things I realized was that my journey was far from over. After a semester to get settled, I taught my first course, a graduate course in quantum mechanics with an emphasis on semiconductor materials and devices. Without question, the only reason I had the faintest idea of how to construct and teach a course was that Dr. Susan Montgomery offered a Teaching Engineering course at the University of Michigan. By far, it was one of the most useful courses I took as a graduate student and the teaching techniques and approaches I learned in that course serve me well to this day. Moreover, while my first semester teaching was a struggle, quantum mechanics has become my absolute favorite course to teach. I know firsthand the pitfalls that can await those new to the subject, and I enjoy sharing my enthusiasm for creating quantum structures in semiconductors (via thin-film deposition and/or nanomaterial synthesis) for use in optoelectronic devices.

As I worked to develop class notes, homework assignments, and exams, for quantum mechanics and other optics-based graduate courses, I also had the prospect of earning tenure looming before me. I set out to distinguish myself from my doctoral advisor and to craft a research program capable of attracting funding and producing publications. My first independent research ideas came from continuing to pursue room temperature, multi-spectral IR photodetectors, the application area in which I had gained expertise as a graduate student. However, I chose to investigate a significantly different material system, namely, colloidal quantum dots (CQDs) embedded in polymer matrices, or hybrid nanocomposites [5]. CQDs are dielectric, semiconductor, or metallic nanoparticles synthesized by inorganic chemistry and surrounded by organic molecules such that they are soluble. I conceived of a unique and distinct approach to demonstrate mid-IR and/or long-wave-IR photodetection in hybrid nanocomposites by embedding CQDs in an electron-conducting polymer matrix to provide a large conduction band offset (or potential well) such that intraband transitions are enabled (Fig. 3.4) [6]. In this way, the conducting polymer not only enhances quantum confinement and electron localization of CQD confined energy levels but also provides photoconduction for electrons that absorb IR light and escape from the CQD. General advantages of CQDs (compared to quantum dots grown by MBE) include: (1) more control over CQD synthesis and the ability to conduct size filtering for highly uniform sizes, (2) the spherical shape of CQDs, which simplifies calculations for device modeling and design, (3) greater selection of light-absorbing materials due to non-existent strain considerations that dominate the growth of EQDs, and (4) the real possibility for room-temperature IR photodetection due to much lower dark currents in hybrid nanocomposites. In fact, my research group demonstrated a room-temperature, mid-IR, intraband spectral response in a hybrid nanocomposite for the first time [7].

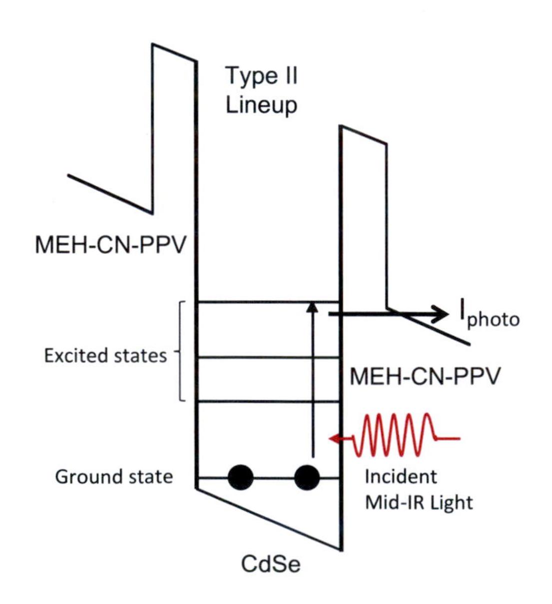

Fig. 3.4 Intraband electronic transitions in the CdSe CQD, surrounded by the conducting polymer barrier, generate photocurrent when subjected to incident mid-infrared radiation. From [6]

However, I needed a new way to deposit thin films of these hybrid materials for the fabrication of IR photodetectors and other optoelectronic devices because the high temperatures used in MBE are not appropriate for large organic molecules. Solution-based processing is the most common deposition technique for hybrid materials; yet, despite its simplicity, it has inherent limitations for hybrid thin films due to the presence of solvent. For example, the optical and electrical properties of organic macromolecules in solution often change when the material is incorporated into a film, and the ability to control these properties is hampered by the variation in structure as solvent evaporates from the film. In addition, it is difficult to deposit blended films of materials having different solubility, and the deposition of multi-layer films with materials having similar solubility is not possible without intermediate blocking layers. In contrast, vacuum-based, dry (i.e., solvent-free) processing, such as physical vapor deposition (PVD), helped usher in the commercialization of organic light-emitting diodes based on small molecules for smart phones, as an example. Specifically, thermal evaporation and sputtering deposition have successfully deposited small organic molecules in device heterostructures; yet, these PVD approaches cannot be used for polymer deposition because the physical processes used to initiate evaporation can destroy long polymer molecular chains.

As a result, in order to deposit polymers (and other macromolecules for hybrid material systems) using PVD, the energy of a given physical process should be absorbed by another medium, or matrix, that coexists in a polymer deposition target, thereby protecting the integrity of the material to be deposited. In addition, in order to meet the promise of hybrid materials, this deposition technique should enable growth of complex heterostructures, provide control of film compositions and nanoscale blending to enable bulk effective media, deposit multi-layer films regardless of solubility, deposit blended films regardless of miscibility, control film morphology (at the surface and within the film bulk), be applicable to a wide range of organic and hybrid thin-film materials, and be compatible with a variety of substrates. For these reasons, I chose to investigate matrix-assisted pulsed laser evaporation (MAPLE), a variation of pulsed laser deposition that can provide dry deposition of hybrid materials (Fig. 3.5) [8]. The MAPLE technique itself is a hybrid approach because the target is solution-based, while the deposition mechanism is a vapor-phase process. Specifically, the solvent in the frozen solution is a matrix that resonantly absorbs the incident laser energy, while the target material is shielded from the incident laser and is transferred from the target to the substrate.

In 2008, my group first introduced a variation of MAPLE known as emulsion-based, resonant infrared matrix-assisted pulsed laser evaporation (RIR-MAPLE) (Fig. 3.6) for organic and hybrid nanocomposite thin films [10]. While MAPLE-related techniques have been under investigation since 1999, emulsion-based RIR-MAPLE is unique because it uses a frozen, complex emulsion as the target [11–13]. An Er:YAG laser (2.94 μm) provides the incident infrared irradiation, which is resonant only with hydroxyl bond vibrational modes. Thus, in the case of polymer and hybrid nanocomposite deposition, the continuous phase of the emulsion is usually water, which serves as the matrix solvent due to the presence of two hydroxyl bonds per molecule and the high absorbance of water ice at the Er:YAG

Fig. 3.5 Photograph of me with the resonant infrared matrix-assisted pulsed laser evaporation (RIR-MAPLE) system in my lab at Duke University circa 2009 (courtesy Pratt School of Engineering, Duke University)

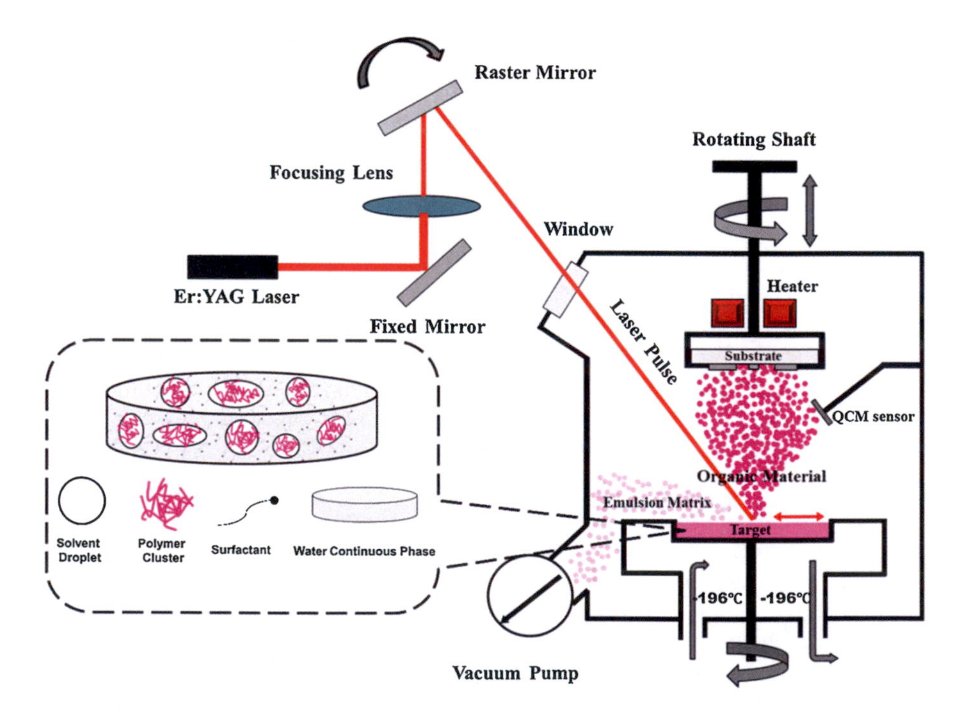

Fig. 3.6 Schematic diagram of an emulsion-based RIR-MAPLE system. The IR Er:YAG laser pulse enters the vacuum chamber through the optical system. The laser irradiates a frozen emulsion target with guest material dispersed inside. The generated material plume is deposited on a substrate, and the emulsion host matrix is vaporized and pumped away. The composition and morphology of the frozen emulsion target are shown in zoomed figure. From [9]

laser energy. The target materials of interest are often hydrophobic and dissolve in organic solvents that serve as the primary solvent and form the oil phase of the emulsion. Importantly, a small amount of surfactant is required to stabilize the oil-in-water emulsion long enough for flash freezing in the target cup before deposition. Finally, phenol is added to the emulsion as a secondary solvent that lowers the overall vapor pressure of the target. During deposition, for each laser pulse, the water matrix evaporates, and due to the low energy of the IR laser source, the target materials are not vaporized or degraded in any way but are transferred intact from the target to the substrate as an emulsified cluster or particle. In this way, the film morphology depends on the emulsified particle size and shape and the order within the emulsified particle.

Thus, emulsion-based RIR-MAPLE essentially decouples the absorption of laser energy from the guest material system, which provides wide flexibility in the choice of organic solvents and target materials. Key to achieving the versatility of this technique is recognizing that the target chemistry in this process must achieve multiple functions: (1) the matrix solvent contains hydroxyl bonds to resonantly absorb the infrared laser energy, (2) the primary solvent dissolves the target material, and (3) the low vapor pressure solvent prevents sublimation of the target under vacuum. Using this principle, my group has used emulsion-based RIR-MAPLE to deposit thin films of hydrophobic polymers [9, 14, 15], hydrophilic polymers [16, 17], metal nanoparticles [18], bulk heterojunctions [19], and hybrid nanocomposites [20, 21]. My group has demonstrated RIR-MAPLE deposition of these materials applied to optical coatings, organic solar cells, infrared photodetectors, gas sensors, and multi-functional surfaces.

Recently, in collaboration with Prof. David Mitzi, my group has started investigating RIR-MAPLE deposition of a very different type of hybrid organic-inorganic material system, namely hybrid perovskites. In contrast to hybrid nanocomposites, hybrid perovskites have an extended crystalline structure that incorporates organic molecules within an inorganic framework. These hybrid materials are exciting because they can provide, simultaneously, the customization of organic molecules and the crystallinity of inorganic materials (i.e., well-defined structures that can be predicted and repeated across large scales). The archetypal hybrid perovskites, such as methylammonium lead triiodide ($MAPbI_3$), feature a three-dimensional crystalline structure comprising corner-sharing metal halide octahedra with a small organic molecule, usually methylammonium, occupying the space created by these octahedra. These three-dimensional hybrid perovskites have demonstrated encouraging device performance as solar cells, reaching device efficiencies on the order of 25% since the first certified demonstration of roughly 13% in 2013 (compared to crystalline silicon solar cells with certified efficiencies around 26% since the first device demonstration in 1954) [22]. In a recent paper with Mitzi, my group reported RIR-MAPLE-deposited $MAPbI_3$ for the first time [23]. In a subsequent manuscript with Mitzi, the first solar cells fabricated using RIR-MAPLE-deposited $MAPbI_3$ films were demonstrated [24]. The best device had stable power conversion efficiency of over 12%, which is comparable to other vapor-phase processed hybrid perovskite solar cells (Fig. 3.7) [25].

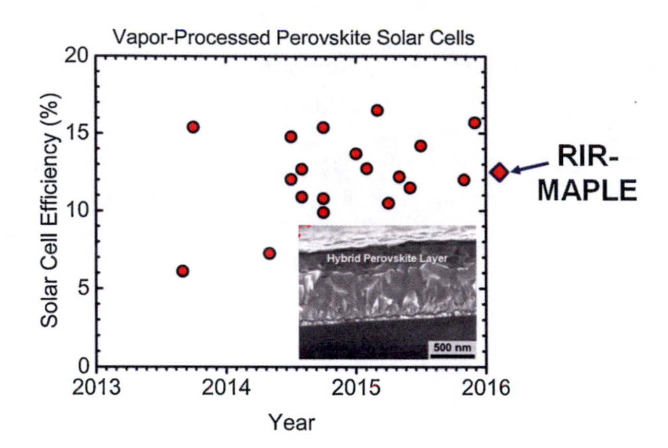

Fig. 3.7 Vapor-processed perovskite solar cell device efficiency [modified from [25] to include RIR-MAPLE hybrid perovskite absorber result of 12.2% power conversion efficiency]. The inset shows the RIR-MAPLE-deposited film

However, one of the most interesting aspects of these materials is that the diversity of hybrid perovskite structures can be increased significantly from the archetypal system by changing the dimensionality of the inorganic framework. By using long, narrow organic molecules as spacers between different numbers of hybrid perovskite sheets (denoted by n), the so-called Ruddlesden-Popper series of hybrid perovskites can be realized. For $n = 2$ and higher, these materials are referred to as quasi-2D because hybrid perovskite sheets are separated by the larger spacer organic molecules. In contrast, the $n = 1$ series comprises layered, two-dimensional perovskites of only the organic spacer molecule and the inorganic metal halide sheet. A common feature of most hybrid perovskites, whether 3D, quasi-2D, or 2D, is that simple organic molecules that are optically or electrically inert are used within the corner sharing octahedra or as the organic spacer molecule. To move beyond $MAPbI_3$ and establish hybrid organic-inorganic perovskites as a new semiconductor technology, there is great potential in using organic molecules that can do more than just provide the required 3D or 2D structure. Instead, using complex, functional organic molecules can enable unique electronic, optoelectronic, or spintronic properties due to synergy between the organic and inorganic components in a crystalline framework.

An important challenge that must be overcome to realize this vision is the synthesis of hybrid perovskite thin films comprising these larger, functional molecules. While thin films are required to enable devices, more complex organic molecules tend to be larger in size and tend to have more complicated chemistries that are not necessarily compatible with the inorganic metal halide for solution-based processing. Meeting this challenge of thin-film synthesis is where my group seeks to contribute to the field. RIR-MAPLE has several benefits for thin-film deposition of hybrid perovskites featuring large, complex organic molecules: (1) the technique offers control of film composition and thickness; (2) gentle deposition is less likely to induce degradation of organic components; (3) solubility problems can be mitigated by using low concentration precursor solutions (~10 mM or less); and (4) perovskite heterostructures of films featuring similar solubility are enabled.

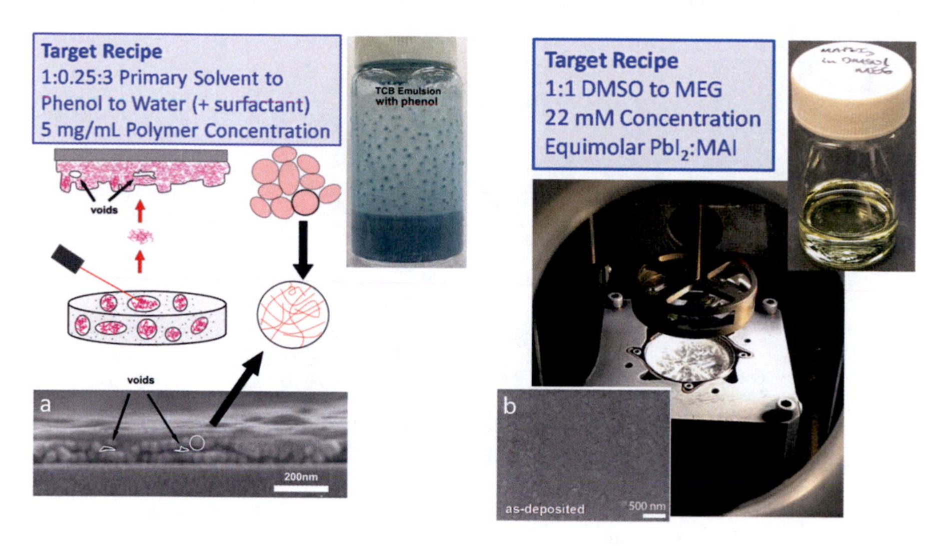

Fig. 3.8 Comparison of RIR-MAPLE deposition of (**a**) polymeric/hybrid nanocomposite thin films and (**b**) hybrid perovskite thin films. While polymeric/hybrid nanocomposite thin films are synthesized by the transfer of emulsified particles from frozen oil-in-water emulsions, hybrid perovskite thin films are formed after organic and inorganic precursors are transferred to the substrate from a frozen solution and annealed to form polycrystalline films

In order to deposit hybrid perovskites using RIR-MAPLE, the target chemistry must be altered to eliminate the presence of water because it degrades hybrid perovskites; yet, water is required for emulsion-based targets (Fig. 3.8). The modified target chemistry has been accomplished by using a polyalcohol, namely monoethylene glycol (MEG) with two hydroxyl bonds, as the matrix solvent and the low vapor pressure solvent. The primary solvent used is dimethyl sulfoxide (DMSO), which is commonly used for perovskite organic and inorganic precursors. With a 1:1 ratio of DMSO to MEG, clear solutions with stoichiometric composition of the organic and inorganic precursors are frozen for use as the target. Upon vaporization of the MEG polyalcohol by resonant absorption of the IR laser energy, the precursor materials in solution are released intact and transferred to the substrate. The perovskite crystal forms, typically due to post-deposition annealing, yielding thin films with crystallinity, morphology, and optical properties that are similar to solution-processed films. The gentle transfer of precursor material, especially in the case of the organic precursors, has enabled the growth of 3D and 2D perovskites, including those with larger, complex, functional organic molecules. As an example, RIR-MAPLE deposition enabled the experimental confirmation of theoretical predictions that tunable quantum well energy level alignments are possible in oligomer-based lead halide perovskites [26]. This work is important because it exemplifies a new paradigm for the design of a wide variety of hybrid perovskites with functional organic molecules. Theoretical predictions of synergistic material properties resulting from interactions of organic and inorganic components can be verified due to the versatility of RIR-MAPLE to deposit a wide range of material systems.

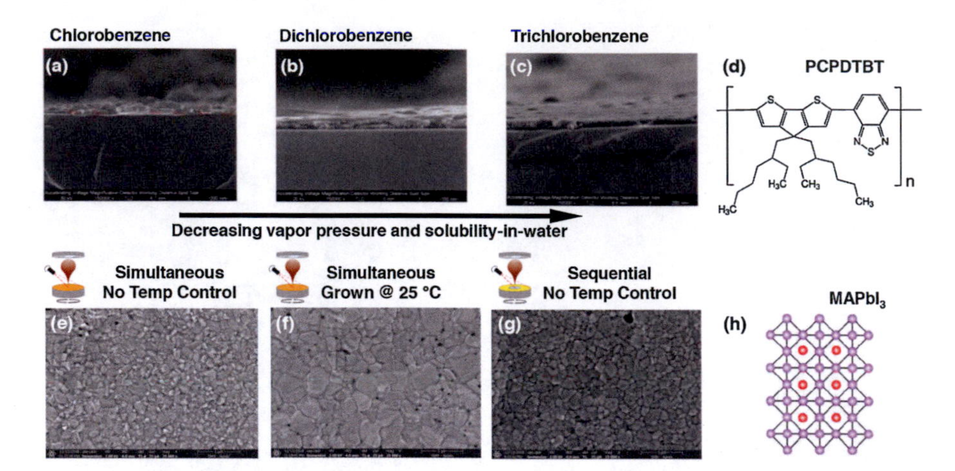

Fig. 3.9 RIR-MAPLE deposition of (**a–d**) PCPDTBT, a polymer used in organic solar cells, demonstrating the dependence of film morphology on emulsified particle size resulting from vapor pressure and solubility-in-water of the primary solvent; and (**e–h**) archetypal hybrid perovskite used in solar cells, $MAPbI_3$, demonstrating the dependence of film morphology on deposition scheme and substrate temperature. RIR-MAPLE growth parameters greatly influence film morphology, and my research seeks to understand the structure-property relationships unique to this deposition technique and the resulting impact on device applications. Part (h) modified from [27]

Over the past 15 years, my research has helped to elucidate the fundamental growth mechanisms of emulsion-based RIR-MAPLE and I have demonstrated its unique suitability for hybrid material systems, ranging from hybrid nanocomposites to hybrid organic-inorganic crystalline semiconductors (Fig. 3.9) [27]. These scientific contributions, along with my earlier work in quantum dot infrared photodetectors, led to my tenure and promotion to Associate Professor in 2011 and promotion to Full Professor in 2018, in each case becoming the first black woman to attain these positions within the Pratt School of Engineering at Duke University. My research efforts have led to experiences never imagined when I first dreamed of becoming a professor in high school. I have had the opportunity to travel across the country and the world and to meet amazing people. I was even fortunate enough to shake hands with President Barack Obama when I received the Presidential Early Career Awards for Scientists and Engineers in 2010. Throughout my research pursuits, with the love and support of family and friends, I have strived to maintain balance in my personal life (most of the time) and to carve out quality time with my husband and two young sons. While research was never the primary motivation that inspired me to earn a PhD, charting new territory and combining physics, electrical engineering, and materials science (with just a little chemistry) to help solve technological challenges, unencumbered by rigid adherence to specific disciplines, has been an exciting and fulfilling addition to the teaching that first attracted me to the profession.

3.4 Responsibility to Help Others

When I consider my academic career, and the ways in which my contributions have a lasting impact, my greatest satisfaction comes from educating the next generations of scientists and engineers. Beyond the typical teaching and research mentorship responsibilities of being a professor, I also serve as the director and lead instructor for the Student Engineers Network: Strengthening Opportunities in Research (SENSOR) Saturday Academy, an education and outreach program that I created for my NSF CAREER award. I designed the SENSOR Saturday Academy to reach underrepresented minority students in the eighth grade with aptitudes for mathematics and science, but little exposure to scientific research as a career. Each year, I recruit a new cohort of approximately 24 eighth grade students to attend 12 sessions of the SENSOR Saturday Academy, plus a field trip, from September to April (Fig. 3.10).

During the SENSOR Saturday Academy, eighth graders learn about different types of sensors used to evaluate water quality. They learn how the sensors work, they practice using the sensors by completing lab experiments, and they learn some of the mathematics that describe sensor operation. The students use the sensors to determine water quality of test samples in the lab. Finally, during a field trip, the students use the same sensors to evaluate water quality during a field to the Eno River State Park in Durham, NC. There are three general 'flavors' of SENSOR Saturday Academy activities. First, the day begins with a lecture that includes one to two chances for group work. The purpose of the lecture is to provide the background knowledge required for the activity of the day. Second, the students participate in a lab, engineering design, or mathematics enrichment activity that uses the information from the lecture and fits in with the overall curriculum to learn how

Fig. 3.10 Photograph of the SENSOR Saturday Academy field trip to the Eno River State Park in Durham, NC in 2018

sensors are used to evaluate water quality. Third, the day ends with an enrichment activity that provides broad exposure to research and life on a university campus. Throughout this academy, students learn about the engineering design process and different types of engineering disciplines.

Thus, the goals of the SENSOR Saturday Academy are: (1) to introduce the engineering design process in the context of testing water quality using sensors (in order to align with the eighth grade science curriculum); (2) to provide meaningful interactions with Duke faculty and graduate students in STEM fields; and (3) to motivate participants in preparation for the transition from middle school to high school by providing academic preparation and by increasing their own expectations. The SENSOR Saturday Academy has been offered for five consecutive years (2014–2019), following a pilot program in 2007. Over this time, a total of 116 eighth graders have been brought into the program. Of these 116 students, 28 students have been invited to return as ninth graders to pursue independent or group engineering design projects, such as 3D printing or the design of solar ovens. A cadre of Duke graduate student mentors makes the program possible by working with the students in small groups to provide additional support and to serve as role models.

It is my great privilege to work with these students, as well as the undergraduate and graduate students that I teach in courses and advise in research, and I always feel invigorated and inspired after interacting with them. When I see a spark ignited in one of my SENSOR students because they learned about a particular engineering discipline, or they made a connection to something they learned at school, or they saw a university lab for the first time, I think back to my early exposure to a university setting and how that has shaped the person I have become. When I teach my university courses, I take the impact that I can have on students' futures very seriously and I create a fair learning environment in which all students can succeed because I know how helpful it was to have instructors with that philosophy in my own education. And when I advise students in research, I try to help them fulfill the unique goals that brought them to graduate school because I know that a PhD is not one size fits all. In my own journey, so many people have contributed to the choices I have made and, often without their knowledge, they have steered me along my path because of seeds they planted that may not have bloomed until many years later. If I can plant that seed for even one student, through the SENSOR Saturday Academy, or a course that I teach, or a student that I advise in research, I will take great pride in knowing that I paid forward all of the opportunities that I have received from others.

References

1. Stiff-Roberts AD (2009) Quantum-dot infrared photodetectors: a review. J Nanophotonics 3:031607
2. Stiff-Roberts AD, Chakrabarti S, Pradhan S, Kochman B, Bhattacharya P (2002) Raster-scan imaging with normal-incidence, midinfrared InAs/GaAs quantum dot infrared photodetectors. Appl Phys Lett 80(18):3265–3267. https://doi.org/10.1063/1.1476387

3. Kochman B et al (2003) Absorption, carrier lifetime, and gain in InAs-GaAs quantum-dot infrared photodetectors. IEEE J Quantum Electron 39(3):459–467. https://doi.org/10.1109/JQE.2002.808169
4. Stiff AD, Krishna S, Bhattacharya P, Kennerly SW (2001) Normal-incidence, high-temperature, mid-infrared, InAs-GaAs vertical quantum-dot infrared photodetector. IEEE J Quantum Electron 37(11):1412–1419. https://doi.org/10.1109/3.958360
5. Stokes EB, Stiff-Roberts AD, Dameron CT (2006) Quantum dots in semiconductor optoelectronic devices. Electrochem Soc Interface 15(4):23–27
6. Lantz KR, Stiff-Roberts AD (2011) Calculation of intraband absorption coefficients in organic/inorganic nanocomposites: effects of colloidal quantum dots surface ligand and dot size. IEEE J Quantum Electron 47(11):1420–1427. https://doi.org/10.1109/JQE.2011.2169235
7. Stiff-Roberts AD, Lantz KR, Pate R (2009) Room-temperature, mid-infrared photodetection in colloidal quantum dot/conjugated polymer hybrid nanocomposites: a new approach to quantum dot infrared photodetectors. J Phys D Appl Phys 42(23):234004. https://doi.org/10.1088/0022-3727/42/23/234004
8. Piqué A et al (1999) Growth of organic thin films by the matrix assisted pulsed laser evaporation (MAPLE) technique. Thin Solid Films 355–356:536–541. https://doi.org/10.1016/S0257-8972(99)00376-X
9. Ge W, Li NK, McCormick RD, Lichtenberg E, Yingling YG, Stiff-Roberts AD (2016) Emulsion-based RIR-MAPLE deposition of conjugated polymers: primary solvent effect and its implications on organic solar cell performance. ACS Appl Mater Interfaces 8(30):19494–19506
10. Pate R, Lantz KR, Stiff-Roberts AD (2008) Tabletop resonant infrared matrix-assisted pulsed laser evaporation of light-emitting organic thin films. IEEE J Sel Top Quantum Electron 14(4):1022–1030. https://doi.org/10.1109/JSTQE.2008.915625
11. Pate R, Stiff-Roberts AD (2009) The impact of laser-target absorption depth on the surface and internal morphology of matrix-assisted pulsed laser evaporated conjugated polymer thin films. Chem Phys Lett 477(4–6):406–410. https://doi.org/10.1016/j.cplett.2009.07.038
12. Pate R, McCormick R, Chen L, Zhou W, Stiff-Roberts AD (2011) RIR-MAPLE deposition of conjugated polymers for application to optoelectronic devices. Appl Phys A 105(3):555–563
13. Stiff-Roberts AD, Ge W (2017) Organic/hybrid thin films deposited by matrix-assisted pulsed laser evaporation (MAPLE). Appl Phys Rev 4(4):041303. https://doi.org/10.1063/1.5000509
14. Ge W, Yu Q, López GP, Stiff-Roberts AD (2014) Antimicrobial oligo (p-phenylene-ethynylene) film deposited by resonant infrared matrix-assisted pulsed laser evaporation. Colloids Surf B: Biointerfaces 116:786–792
15. McCormick RD, Cline ED, Chadha AS, Zhou W, Stiff-Roberts AD (2013) Tuning the refractive index of homopolymer blends by controlling nanoscale domain size via RIR-MAPLE deposition. Macromol Chem Phys 214(23):2643–2650. https://doi.org/10.1002/macp.201300465
16. Yu Q, Ge W, Atewologun A, López GP, Stiff-Roberts AD (2014) RIR-MAPLE deposition of multifunctional films combining biocidal and fouling release properties. J Mater Chem B 2(27):4371–4378
17. Yu Q, Ge W, Atewologun A, Stiff-Roberts AD, López GP (2015) Antimicrobial and bacteria-releasing multifunctional surfaces: oligo (p-phenylene-ethynylene)/poly (N-isopropylacrylamide) films deposited by RIR-MAPLE. Colloids Surf B: Biointerfaces 126:328–334
18. Ge W, Hoang TB, Mikkelsen MH, Stiff-Roberts AD (2016) RIR-MAPLE deposition of plasmonic silver nanoparticles. Applied Physics A 122(9):824
19. Ge W, McCormick RD, Nyikayaramba G, Stiff-Roberts AD (2014) Bulk heterojunction PCPDTBT: PC71BM organic solar cells deposited by emulsion-based, resonant infrared matrix-assisted pulsed laser evaporation. Appl Phys Lett 104(22):223901
20. Pate R, Lantz KR, Stiff-Roberts AD (2009) Resonant infrared matrix-assisted pulsed laser evaporation of CdSe colloidal quantum dot/poly[2-methoxy-5-(2′-ethylhexyloxy)-1,4-(1-cyano vinylene)phenylene] hybrid nanocomposite thin films. Thin Solid Films 517(24):6798–6802. https://doi.org/10.1016/j.tsf.2009.06.018

21. Ge W, Atewologun A, Stiff-Roberts AD (2015) Hybrid nanocomposite thin films deposited by emulsion-based resonant infrared matrix-assisted pulsed laser evaporation for photovoltaic applications. Org Electron 22:98–107
22. Best research-cell efficiency chart. Photovoltaic Research.NREL.https://www.nrel.gov/pv/cell-efficiency.html.Accessed 15 Aug 2019
23. Barraza ET, Dunlap-Shohl WA, Mitzi DB, Stiff-Roberts AD (2018) Deposition of methylammonium lead triiodide by resonant infrared matrix-assisted pulsed laser evaporation. J Electron Mater 47:917. https://doi.org/10.1007/s11664-017-5814-0
24. Dunlap-Shohl WA, Barraza ET, Barrette A, Gundogdu K, Stiff-Roberts AD, Mitzi DB (2018) MAPbI3 solar cells with absorber deposited by resonant infrared matrix-assisted pulsed laser evaporation. ACS Energy Lett 3(2):270–275. https://doi.org/10.1021/acsenergylett.7b01144
25. Ono LK, Leyden MR, Wang S, Qi Y (2016) Organometal halide perovskite thin films and solar cells by vapor deposition. J Mater Chem A 4(18):6693–6713. https://doi.org/10.1039/C5TA08963H
26. Dunlap-Shohl WA et al (2019) Tunable internal quantum well alignment in rationally designed oligomer-based perovskite films deposited by resonant infrared matrix-assisted pulsed laser evaporation. Mater Horiz 6:1707–1716. https://doi.org/10.1039/C9MH00366E
27. Dunlap-Shohl WA, Zhou Y, Padture NP, Mitzi DB (2019) Synthetic approaches for halide perovskite thin films. Chem Rev 119(5):3193–3295. https://doi.org/10.1021/acs.chemrev.8b00318

Chapter 4
Nanoscale Materials Engineering for Microelectronics

Santosh K. Kurinec

4.1 Introduction and Background

4.1.1 The Foundations

I grew up in New Delhi, India in a middle class diverse neighborhood designed by the British for mid-level government employees. After middle school, English was the medium of instruction in the school I attended. India is a diverse country with over 22 languages. English and math connected us. Math and science curricula were being heavily emphasized and invested in, as the Indian government was seeking to lift itself to newer heights in a post-colonial era. I reason that why my parents cultivated an environment at home that preached education as the ultimate path to self-determination.

Fast forward to the 1960s, when the education branch of the Indian government created the National Science Talent Search Scheme, or NSTSS. This nationwide program had the purpose of giving scholarships to motivate students to pursue higher education in science.

When I reached high school, my teachers encouraged me to apply to the NSTSS. The application consisted of submitting a science project report of my choosing, in additional to an aptitude test, written essay, and a final interview. My project was a study on dairy fermentation and how lactic acid is formed as milk becomes yogurt. I set up a titration system at home to determine acidity as a function of time, temperature, and dopants such as salt and sugar. While challenging at the time, in hindsight I fondly remember sharing the tight quarters in our kitchen with my mother, plotting pH charts by hand on occasionally milk-soaked sheets of graph paper.

S. K. Kurinec (✉)
Electrical and Microelectronic Engineering, Rochester Institute of Technology, Rochester, NY, USA
e-mail: skkemc@rit.edu

A. C. Parker, L. Lunardi (eds.), *Women in Microelectronics*, Women in Engineering and Science, https://doi.org/10.1007/978-3-030-46377-9_4

My project report would go on to be accepted in the program, and following a successful application process, I was ultimately selected as one of the winning finalists! While the recognition itself was a huge confidence boost (and an excuse to celebrate with family and friends), it was the scholarship that really made the difference. I decided to join Bachelor of Science program in Physics at the University of Delhi. As a young student, the first connection in our home to the outside world was a Philips tube radio. In the sweltering heat of July 1969, I will never forget sitting by that radio and hearing those barely audible words crack through the speakers. "The Eagle has landed!" That was a stunning excitement and took our imaginations beyond belief. My interest in science was further heightened.

The NSTSS program mandated summer internships each year up to the Masters level at leading research organizations and universities. I attended five summer schools—at Punjab University, Chandigarh, Indian Institute of Technology, Delhi, Indian Institute of Technology, Kanpur, Bhabha Atomic Energy Research Center, Bombay, and Tata Institute of Fundamental Research, Trombay. During these internships, I learnt using X-ray diffraction, neutron scattering, and nuclear magnetic resonance techniques for advanced materials/chemical analysis. I graduated with MS in Physics. I had several job opportunities but decided to pursue PhD. Without the NSTSS scholarship, it is unlikely that I could have pursued education to the doctoral level and eventually make it here today.

4.1.2 PhD Research

I joined the National Physical Laboratory (NPL), New Delhi for my PhD research. The National Physical Laboratory is the premier research laboratory in India in the field of physical sciences. India was exploring development of magnetic materials for booming electrical, electronics, and automotive industries using indigenous raw materials. However, iron oxide, the major raw material extracted from Indian ores had silicon impurity whose effect on magnetic properties was not well understood. I took it as my PhD research topic and studied the effect of dopants like Si, Ge, and Sn on the permeability of sintered ceramic Mn-Zn ferrites. While most of the efforts were done to optimize sintering parameters for desired characteristics, the NPL group at that time did not explore the effect of microstructure on magnetic properties. I learnt that Indian Institute of Technology just received the first Cambridge scanning electron microscope and I started to investigate the microstructure of ferrites as a function of doping and sintering conditions. I discovered a relationship between the grain size and porosity on the magnetic properties. How do dopants affect the microstructure needed further accurate materials analysis. I learnt about Auger Electron Spectroscopy (AES) from a speaker from Physical Electronics (PHI)—a newly founded spin-off company from the University of Minnesota. I analyzed solubility and grain boundary segregation of Si, Ge, and Sn in $Mn_{1-x}Zn_{x}Fe_2O_4$ using SEM, XPS, and AES. These analyses led us to form low loss, high permeability ($\mu_r \sim 10{,}000$) Mn-Zn ferrites with relatively impure raw materials.

India today is one of the major ferrite producing countries. I also investigated conduction mechanism in ceramic lanthanum strontium chromate for high temperature applications. I received my doctorate degree from Ms. Indira Gandhi—one of the first woman Prime Ministers in history, and chancellor of the University of Delhi.

4.1.3 Scientist—National Physical Laboratory

The energy crisis of the early eighties brought me to the area of Photovoltaics (PV). I was then hired by the NPL as a Scientist in the Division of Materials to work on the development of low cost polysilicon solar cells. We had an in-house process for refining metallurgical grade silicon and then zone refining polysilicon rods. The silicon we were using had significant iron impurity. My research was focused on the study of Fe in Si, inverse of what I studied in my PhD work. Our group developed the process for making solar cells using $POCl_3$ emitter diffusion and Al-Ag contacts yielding 12% efficient solar cells. Based on my PV work, I received support from the University of Florida to participate at the Training in Alternative Energy Technologies (TAET) program.

4.1.4 Moving to the USA

I joined the Materials Science and Engineering department of University of Florida (UF) in Professor Holloway's group to work on Ni-Cu metallization for electronic packaging. I was also operating X-ray photoelectron spectroscopy (XPS) System at UF Major Analytical Instrumentation Center (MAIC). In addition, I continued working on investigating the effect of diffusion in the grain boundaries in polysilicon solar cells. After about year and half postdoc, I received faculty job at Florida A&M/Florida State University College of Engineering as Assistant Professor of Electrical Engineering. The Strategic Defense Initiative Organization (SDIO) was set up in 1984 within the US Department of Defense to support research in the fields of high-energy physics, optical communications, supercomputing/computation, and advanced materials. I started working on light emission from silicon using reverse biased diodes and newly observed light emission from porous silicon. I also used my ceramic experience to deposit superconducting yttrium barium copper oxide thin films.

That was the time when CMOS was picking up at submicron node. That brought me to Rochester Institute of Technology (RIT) in 1988 as Associate Professor of Microelectronic Engineering. RIT had founded a unique Bachelor of Science program in Microelectronic Engineering in 1982. I brought my materials science perspective to this program developing new courses, processes, and capabilities. Today the program is supported with a complete 6-in. CMOS line equipped with advanced lithography tools, thermal diffusion, ion implantation, plasma, CVD and ALD processes, chemical mechanical planarization (CMP) and device design, modeling and test laboratories. Over 1500 alumni from this program are contributing to the semiconductor industry worldwide.

4.2 Microelectronics

4.2.1 Emergence of Microelectronics

Figure 4.1 shows the electronic device technology that went on the moon in 1969, bipolar transistor-based logic, ferrite magnetic core memory and GaAs solar cells. It was the first monolithically integrated bipolar chip made by Fairchild.

That was the beginning of the semiconductor technology moving to metal oxide semiconductor field effect transistors (MOSFETs). From p-channel to n-channel and then to complementary metal oxide semiconductor field effect transistor (CMOS) technology. With Gordon Moore predicting transistor scaling in 1965, the technology has continued with unprecedented growth from micron scales to nanometer scales. With each technology node, new material innovations have emerged in modern CMOS. From aluminum gate/SiO_2 gate dielectric/Si MOS, the technology has brought in new dielectrics and electrodes and integrated them at nanoscales to achieve higher performance and speed. While materials have been implemented,

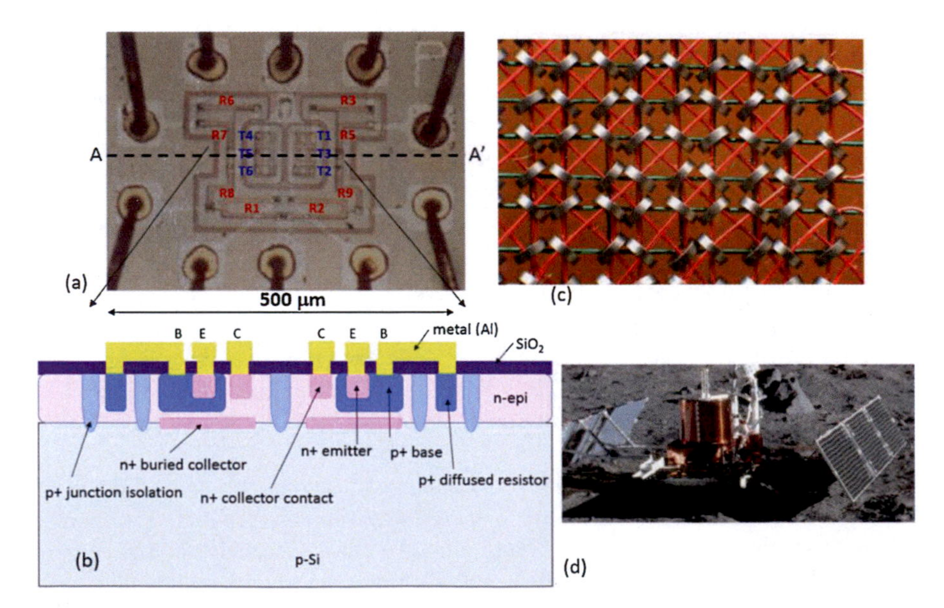

Fig. 4.1 Technologies used in the Apollo Guidance Computer in 1969. (**a**) 3 input NOR gate (**b**) Cross section across AA'; (**c**) ferrite core memory, and (**d**) GaAs solar cell panel [(**a**, **c**, **d**) Wikipedia]

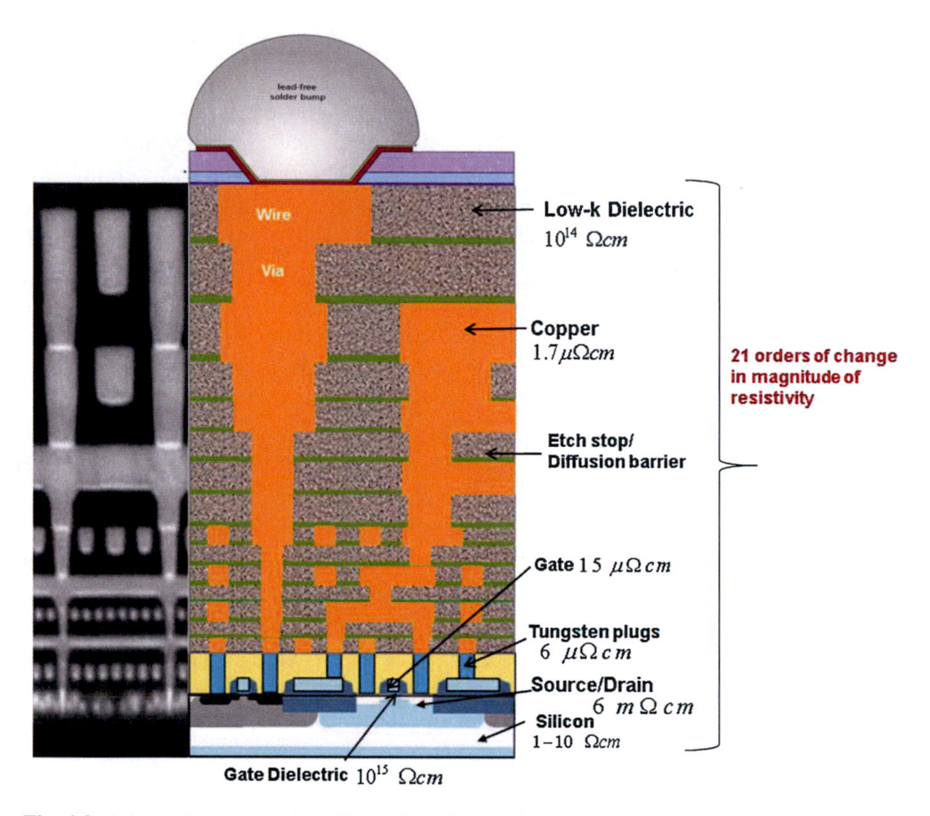

Fig. 4.2 Schematic cross section of a semiconductor chip showing resistivity variation in different layers. (Recreated from sources: Intel Development Forum IDF2012, ITRS 2011)

novel methods of deposition, patterning, etching, and analysis have evolved. In the modern chip, as demonstrated in Fig. 4.2, the conductivity varies by ~21 orders of magnitude over a volume of the order of 1 μm^3!

4.2.2 Scaling of CMOS: Planar to Non-Planar, New Materials, and Processes

Development of smaller, faster, cheaper devices will continue pushing faster clock speeds on microprocessors and increasing storage capability in memory devices. Manufacturing innovations have demonstrated readiness to implement 10 nm and 7 nm nodes on 300 mm and possibly on 450 mm wafers. New process technologies—nanolithography (i-line, immersion, EUV, directed self-assembly (DSA)), atomic layer deposition and etching, plasma doping, spike anneals, chemical mechanical planarization, etc.; substrate engineering (strained Si, SOI, SiGeOI), new materials (high and low k dielectrics, silicides, ferroelectrics and magnetics,

2D semiconductors), novel packaging (Flip Chip, RDL, 3D TSV), and software developments have exceeded predictions.

With continued scaling of the MOSFETs, non-ideal effects such as short channel effects (SCEs), poor electrostatics integrity, and large device variability start to appear. The transistor structure has radically changed from its original planar 2D architecture to modern 3D Fin field-effect transistors (FinFETs) along with new designs for gate and source/drain regions and applying strain engineering. Currently, bulk FinFETs have been widely used in mass production from 22 nm to the 10 nm node and will be extended to the 5 nm node. At the end of the technology roadmap, the 3 nm node and the traditional bulk FinFET technologies would encounter enormous challenges. Thus, new device structures, new materials, and new integration approaches are being explored. Therefore, novel promising device architectures like fin-on-insulator (FOI) FinFET, stacked gate-all-around (GAA) NW, and nanosheet are receiving increasing attention. Other materials such as 2D graphene, MoS_2, and carbon nanotube are under intense research to replace and/or augment silicon (Fig. 4.3).

With my experience in magnetic, optical, and electronic materials, I focused my research on integrating novel materials and devices onto CMOS platform for heterogeneous integration. This required inter-disciplinary, inter-institutional, and global collaborations.

The fundamental scientific and technological properties are related to the chemical structure of a material at the nanometer or even atomistic length scales. This includes but not limited to, internal interfaces of complex topology in devices such as advanced nanoelectronic and quantum structures and metallic alloys. In modern

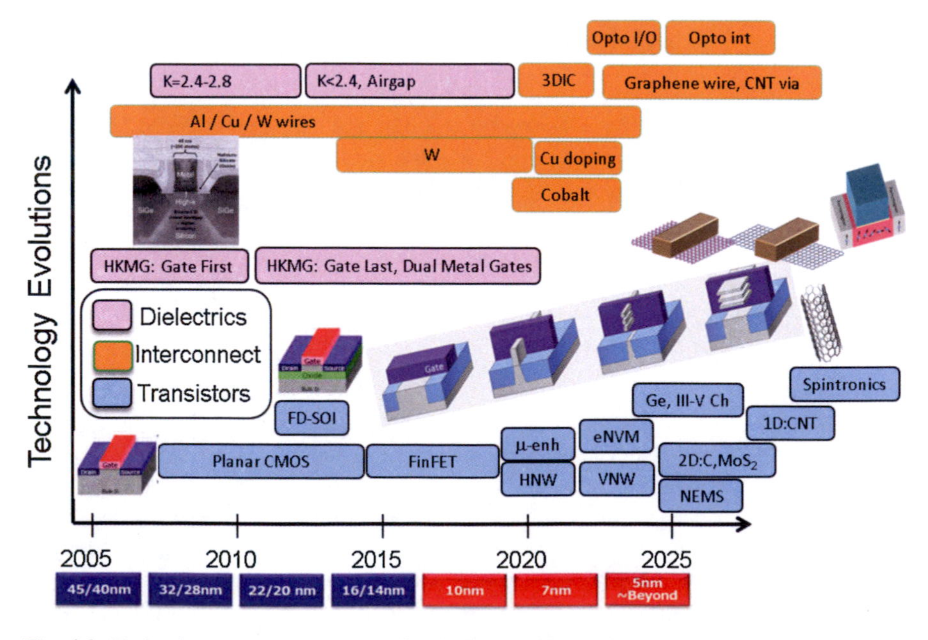

Fig. 4.3 Technology evolutions with scaling in the CMOS transistors

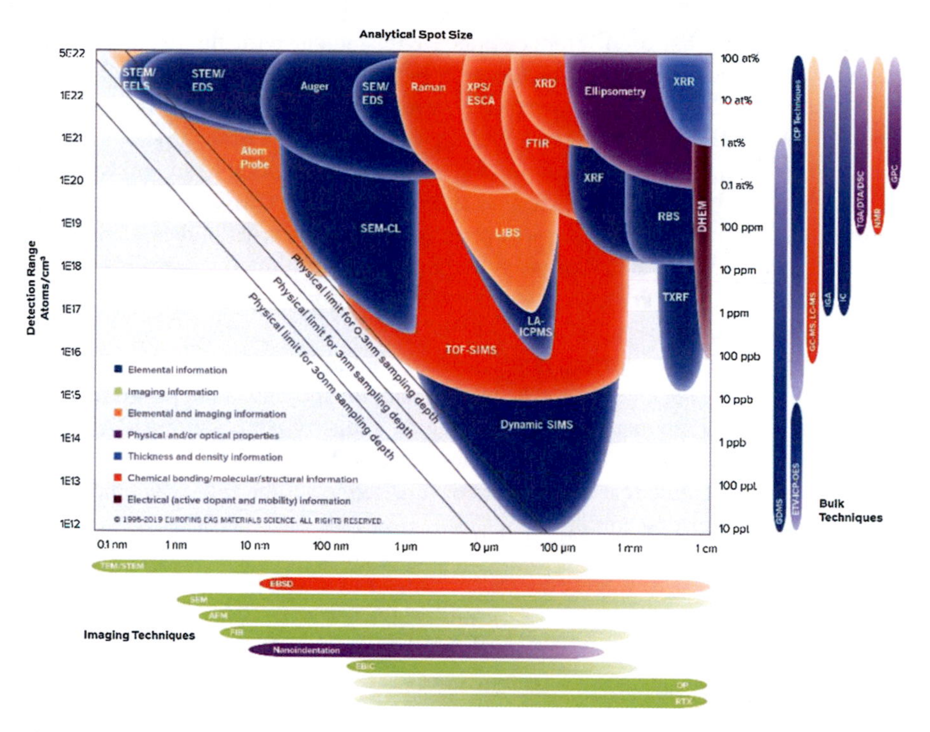

Fig. 4.4 Detection limits and analytical resolutions of various techniques developed over the years used for materials characterization. (Included with permission; copyright© EAG Materials Science, LLC, https://www.eag.com/techniques/)

devices, distribution of a few atoms can determine the performance of the device. Breakthroughs in materials science are often concomitant to advances in characterization methods. Figure 4.4 shows the EAG SMART (Spectroscopy and Microscopy Analytical Resolution Tool) chart that provides a concise visual reference for comparing analytical techniques. Development of these techniques has been the key in engineering materials at nanoscales. I have employed most of the techniques in my research, currently exploring Atomic Probe Tomography (APT) in monolayer doping.

> If you cannot measure it, you cannot improve it—Lord Kelvin
>
> If you can see it, you can make it—Santosh Kurinec

4.2.3 *Microelectronic Engineering Education*

I strongly believe in integrating research in undergraduate education, particularly in the continuously evolving field of microelectronics. My teaching methods span the range from lectures, labs to projects. In performing the capstone projects, students conduct research that often produces deep engagement with the team of faculty, technical staff and students and often with industrial partners. I have advised over 170 such projects in the last 25 years that have resulted in the development of modern microsystem fabrication processes in the RIT lab. These processes, rarely encountered in undergraduate educational experiences, include hands-on work with metal gates, high-k gate dielectrics, silicides, non-planar CMOS, magnetic tunnel and resonant tunnel devices, low-k copper, MEMS, carbon nanotubes, non-volatile memory, and photovoltaics. The program requires at least three semesters of internship that has been as a great way of generating high quality workforce for the semiconductor industry.

With major NSF Department Level Reform funding, I designed and implemented an innovative state-of-the-art curriculum in nanotechnology. This program helped my department to interact with Semiconductor Industry Association (SIA), IMEC Belgium and outreaches to K-12 teachers. Generation of Microelectronic Engineering workforce that can address the challenges of the future beyond CMOS is essential and must be continued.

4.3 Microelectronic Materials and Devices Research

My research has been primarily on applied materials science with a vision on applications. How the properties can be used to create devices with desired output characteristics that have the potential to be manufactured into products (Fig. 4.5).

4.3.1 Magnetics

4.3.1.1 Cu-Ferrite Inductor

With my ferrite experience, I initiated a collaboration with a company, Inframat Corporation, and Professor Jayanti Venkataraman from RIT Electrical Engineering to fabricate ferrite inductor on a chip. Inductive components are extensively used in high frequency (>1 MHz) electronic devices from radar, satellite, telecommunication systems to home radio stereos. Conventional inductive components use metallic alloys and ferrites as core materials. The major problem for metallic materials is their low resistivity (~10^{-6} Ω-cm). Since it is impossible to dramatically increase their resistivity, metallic materials were excluded in high frequency applications and ferrites have been the choice for decades. Electrophoretic deposition (EPD) of ferrite nanoparticles was employed to fill in the on-chip copper coils (Fig. 4.6).

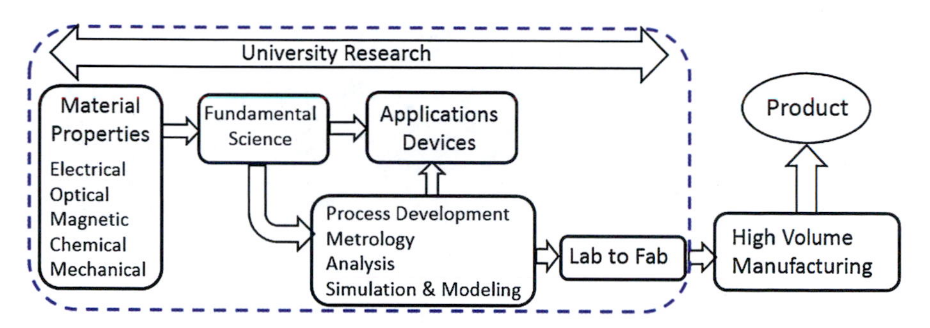

Fig. 4.5 Research strategies in integration of novel materials in practical devices

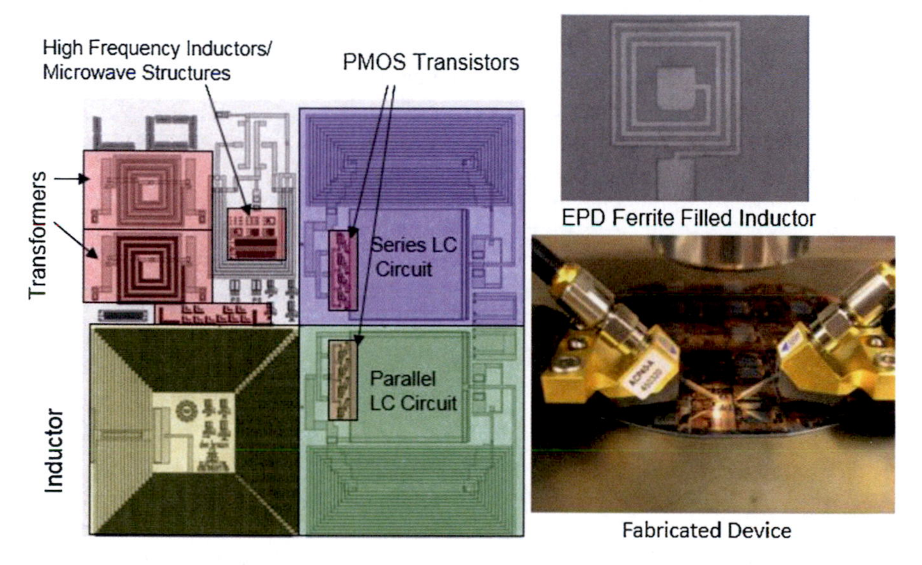

Fig. 4.6 Electrophoretically deposited ferrite inductor integrated with copper coils on silicon wafer

4.3.1.2 Magnetic Tunnel Junctions

Magnetic tunnel junctions with a MgO tunneling layer have been used in memory devices after theoretical predictions of significantly high tunneling magnetoresistance (TMR) were reported. TMRs of 500% at room temperature and 1010% at 5 K for CoFeB|MgO|CoFeB structures have been achieved. The sputter deposited CoFeB film is amorphous and subsequently crystallizes on annealing using MgO as a template for forming a continuous (100) out-of-plane oriented crystal structure, which is critical to obtain high TMR. It was well known that the diffusion of B out of the CoFeB is necessary for proper crystallization of the CoFeB. However, boron diffusion was not well studied, as the layers are very thin to provide enough signal in XRD/TEM/EELS analysis. My PhD student then, Sankha Mukherjee (now at

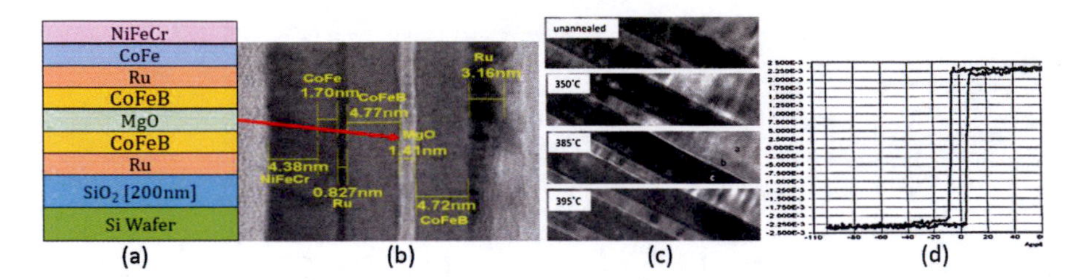

Fig. 4.7 (**a**) Schematic of the MTJ stack integrated on Si; (**b**) TEM micrograph of the structure annealed labelled based on the PEELS (parallel electron energy loss spectroscopy) analysis; (**c**) effect of annealing temperature; (**d**) Magnetization curves using vibrating sample magnetometer (VSM)

Holmusk, Singapore) fabricated several layer stacks to investigate boron diffusion and CoFeB crystallization (Fig. 4.7).

Subsequently we developed a low temperature inter level dielectric to integrate MTJ on silicon wafers and also developed MTJ SPICE model. This work could not have been done without the support from Hitachi, Veeco Instruments and Micron and funding from the National Science Foundation.

4.3.2 Optical Materials and Devices

The Phosphor Technology Center of Excellence was established at the Georgia Institute of Technology in 1994 with other university members. The research mission addressed needs in five technological areas: cathode ray tube, electroluminescence, field emission devices, plasma display panels, and active-matrix liquid crystal display back-light phosphors through interactive university/industry technology groups. Inspired by this, we started research on phosphors in collaboration with a local start-up, Advanced Vision Technologies, Inc. We observed blue electroluminescence by reacting ZnO with Ta_2O_5 and identified the phase as $Ta_2Zn_3O_8$ (FCC space group C2/c, PDF # 521764). It was also realized in thin film form on silicon substrate, making TZO an important material for monolithic field emission display devices. Using electrophoretic deposition and photolithography a three-color phosphor screen process was developed in collaboration with Esther Sluzky at Hughes Aircraft. My thin film metallic alloy experience came useful in investigating aluminum films for digital light processing (DLP) displays (Fig. 4.8).

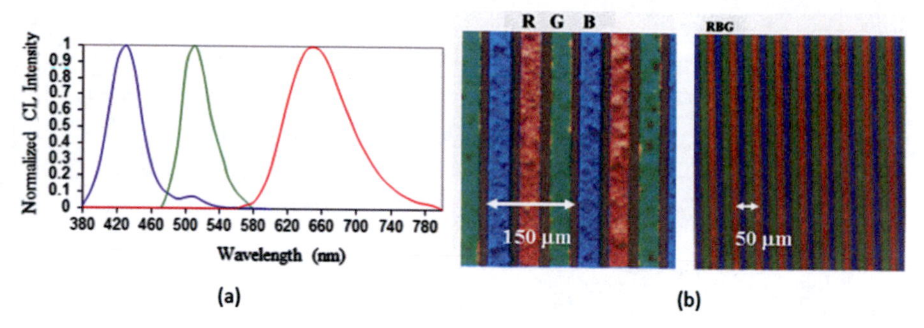

Fig. 4.8 (**a**) Blue Cathodoluminescence from $Ta_2Zn_3O_8$; green and red from Mn doped TZO; (**b**) Three-color phosphor screen fabricated using electrophoretic deposition and photolithography (*US Patent 5,582,703*, December 1996)

4.3.3 Electronic Devices

4.3.3.1 Resonant Tunnel Diodes

When I joined RIT, it did not have a PhD program. Most of my research was conducted with undergraduate students through their capstone projects. Subsequently, Masters Programs were developed and then the first PhD program in Microsystems Engineering in the year 2002. I involved my first PhD student (Stephen Sudirgo, now at Intel) to explore integration of resonant tunneling diodes (RTDs) with CMOS. RTDs are promising band gap engineered heterostructures that exhibit a distinctive negative differential resistance (NDR) that can be exploited for various electronic functions. A strong collaboration with Professor Paul Berger from the University of Delaware (now at Ohio State University) and his PhD student Prof. Sean Rommel, now at RIT as the Director of the Microelectronic Engineering programs, was established. Among the integration strategies investigated, placing of tunnel diodes on the source/drain regions of MOS transistors was found to be most promising. The proposed goal was to minimize this impact and limit the number of steps in the process (Fig. 4.9).

Integration was by growing the Si-based tunnel diode onto pre-existing CMOS circuitry following all high temperature CMOS processing but prior to CMOS metallization.

4.3.3.2 Emerging Semiconductor Memory

At no time in the history of semiconductor industry has memory technology assumed such a pivotal position. The last decade has seen a remarkable shift in usage and value of semiconductor memory technologies. These changes are driven by growth of mobile multi-media applications, sheer volume of data, and need for subsystems and multi-die packages rather than as discrete components on a motherboard.

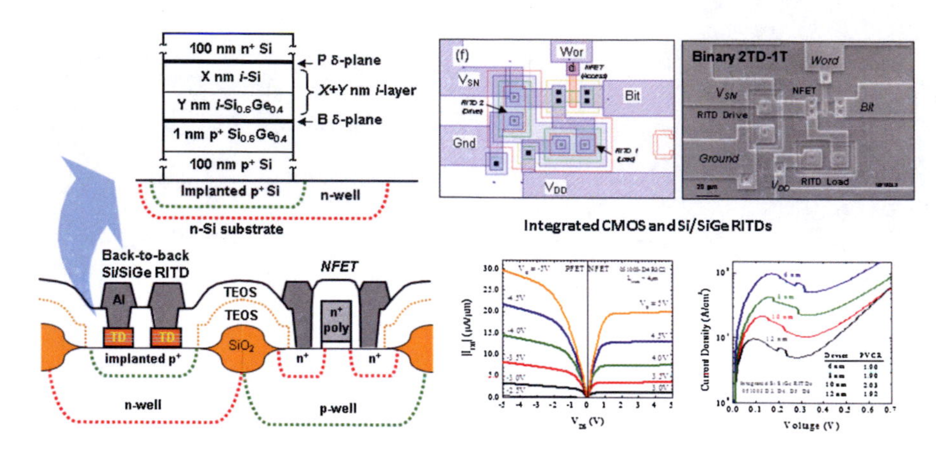

Fig. 4.9 Si based RTD integrated on CMOS platform

Over the past three decades, numerous memory technologies have been brought to market, for example: **S**tatic **R**andom-**A**ccess **M**emory (SRAM), Pseudo Static RAM, NOR Flash, **E**rasable **P**rogrammable **R**ead-**O**nly **M**emory (EPROM), **E**lectrically **E**rasable **P**rogrammable **R**ead-**O**nly **M**emory (EEPROM), Dynamic RAM (DRAM), and NAND flash. Generally speaking, these memory technologies can be split into two categories, volatile and non-volatile; volatile memory does not retain data when power is turned off, conversely, non-volatile memory retains data once power is turned off. The dominating memory technologies in the industry today are SRAM, DRAM (volatile), and NAND flash (non-volatile). Storage class memory (SCM) describes a device category that combines the benefits of solid-state memory, such as high performance and robustness, with the archival capabilities and low cost per bit of conventional hard disk magnetic storage. Such a device requires a non-volatile memory technology that can be manufactured at a very low cost per bit (Fig. 4.10).

In the past decade, significant focus has been put on the emerging memories field to find possible contenders to displace either or both NAND flash and DRAM. Some of these emerging technologies include: MRAM (Magnetic RAM), STT-RAM (Spin-Transfer Torque RAM), FeRAM (Ferroelectric RAM), PCRAM (Phase Change RAM), RRAM (Resistive RAM), and Memristor based RRAM. Each of these relies on either charge transfer or resistance change via different mechanisms. Currently, any competing solid-state memory technology has to either outperform flash memory in its own memory segment—which is difficult in terms of density unless multi-bit per cell operation is achieved or has to offer higher performance. One approach to attain high-density, high-throughput, and low cost memory system is to pursue 3D integration of logic and memory operating in unison.

In 2009–2010, I took academic sabbatical at IBM T.J. Watson Research Center. I initiated collaboration with Dr. Simone Raoux (now Professor, *Helmholtz-Zentrum Berlin für Materialien und Energie*, *Germany*) on investigating phase change materials (PCM). I with my PhD student Archana Devasia (now at NASA

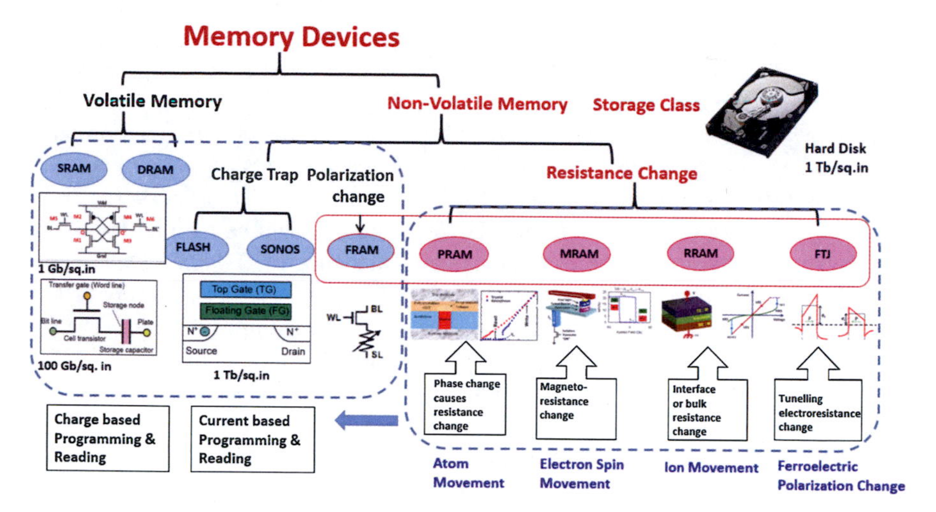

Fig. 4.10 Classifications of memory devices (Recreated from sources: Leti, 2012, Yole Development, 2017, France)

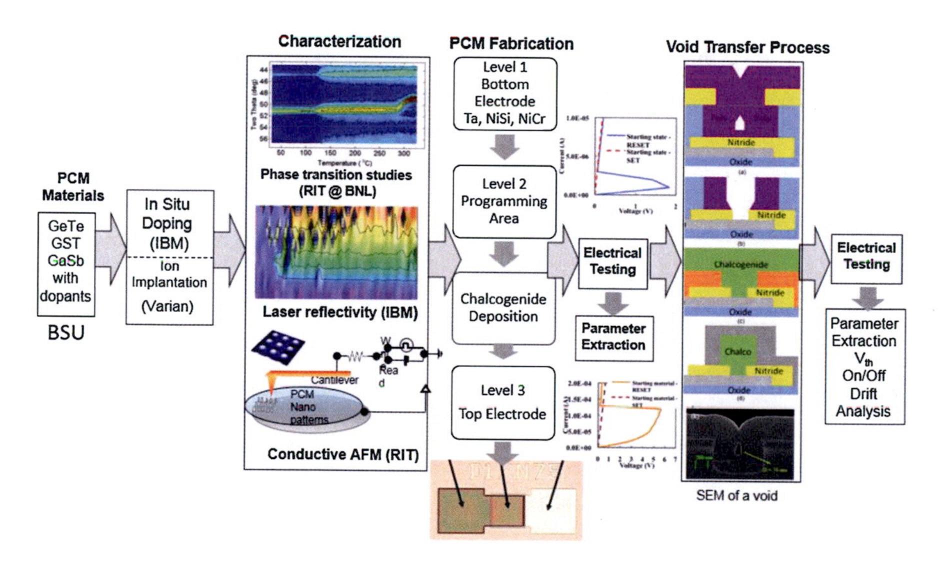

Fig. 4.11 Inter-university–industry collaborative research on PCM materials analysis for threshold voltage and switching characteristics

Goddard) was already collaborating with Professor Kris Campbell of Boise State University on bilayer chalcogenide PCM stacks. With IBM support, we got access to IBM X-20 line at National Synchrotron Light Source (NSLS) at Brookhaven National Laboratory to study time resolved X-ray diffraction on PCM materials (Fig. 4.11).

The discovery of ferroelectric behavior in HfO_2 by Boscke et al. (Appl. Phys. Lett. 99, 102,903 (*2011*); at NaMLab, Germany has led to tremendous growth in research on ferroelectric films and devices. HfO_2 exhibits a range of ferroelectric properties based on the dopants added and the anneal process that is performed. It is CMOS compatible and is scalable. I have developed collaboration with NaMLab and currently researching on Ferroelectric Field Effect Transistors (FeFETs). Ferroelectric high dielectric constant (High-k) material can be applied for a more complicated transistor design, e.g., negative capacitance FET (NCFET). This type of transistor is a strong potential device beyond the 5 nm node CMOS (Fig. 4.12).

The reason for choosing NCFET is due to its dramatic improvement in the sub-threshold swing, which has compatible process flow with the conventional CMOS technology, and on-current enhancement. The high-k materials suitable for NCFET are considered to be $HfZrO_2$ and HfO_2 with a thickness below 5 nm.

Thin ferroelectrics-based tunnel junctions have the potential as a memristive device based on tunneling electro resistance. The dipoles in ferroelectric film switch up or down depending on voltage polarity. If the thickness of the ferroelectric layer is made small enough, it can allow tunneling of electrons that is a function of the relative density of dipoles in up or down position thus preserving a memory similar to that of the synapse. Creating this device will enable making electronic analog circuits to mimic the brain which currently being investigated in collaboration with Professor Nathaniel Cady, SUNY Polytechnic and Professor Dhireesha Kudithipudi, now at University of Texas, San Antonio (Fig. 4.13).

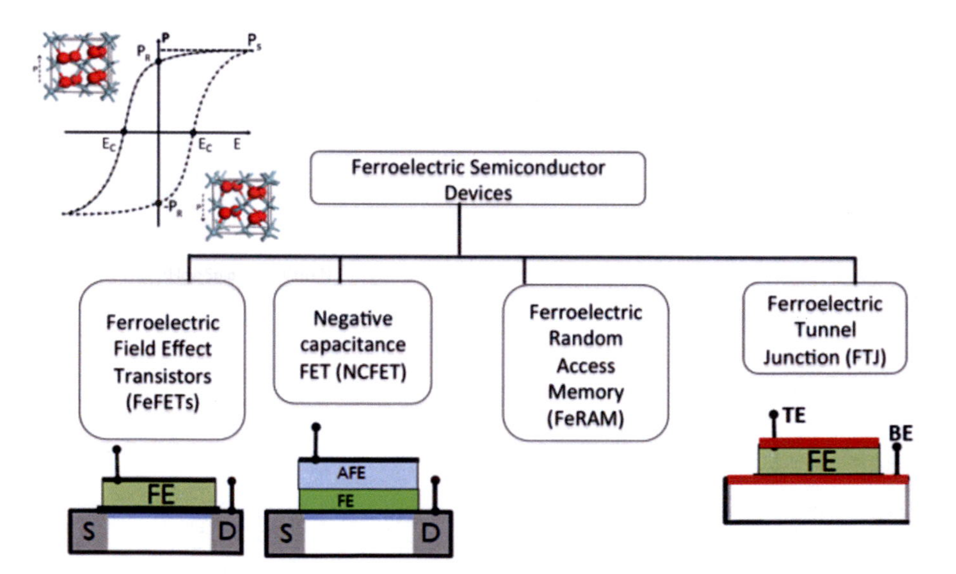

Fig. 4.12 Emerging frontiers of ferroelectric semiconductor devices

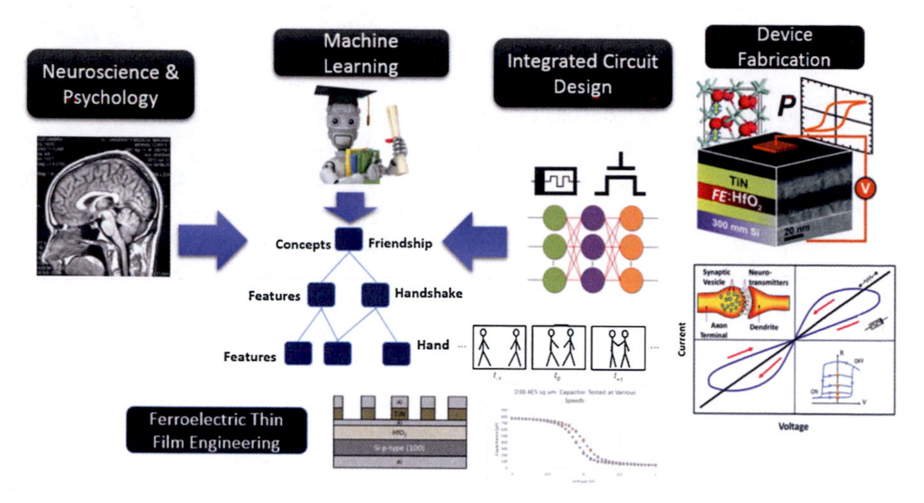

Fig. 4.13 Schematic of a brain inspired machine learning system utilizing ferroelectric memristor devices

4.3.3.3 Other Research Areas

As the technology is constantly evolving, new processes and materials are being investigated. I studied MoS_2 under high resolution TEM and EELS to examine the layers structure. For conformal ultra-shallow junction formation, my group is currently working on developing monolayer doping (MLD). We have fabricated diodes and transistors (FETs) using MLD (Fig. 4.14).

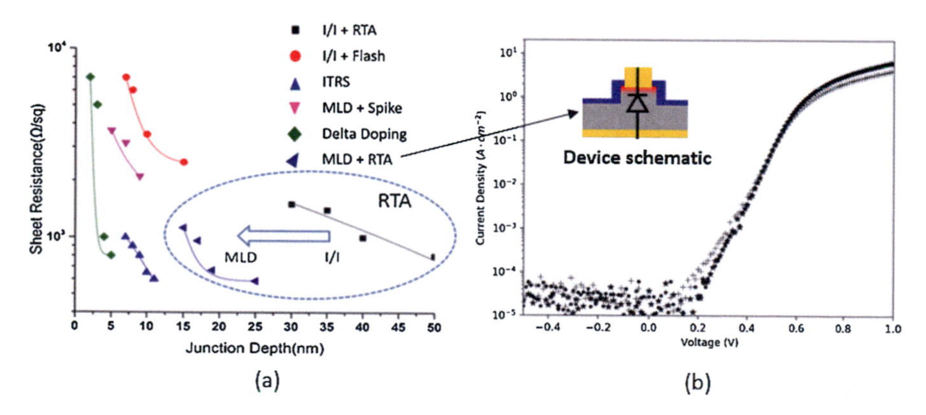

Fig. 4.14 (a) Sheet resistance—junction depth (defined at 1E18 cm^{-3} substrate doping) map for different doping technologies. MLD shows lower values of sheet resistance and junction depth compared to other techniques; (b) I–V characteristics obtained from MLD doped diodes

4.4 Conclusions

My research has spanned across electronic and magnetic materials in bulk as well as in thin films. When we move from one technology to the other, we bring experience and new perspective. Working on magnetic and ferroelectric ceramics motivated me to work on MTJs and FeFETs. The photovoltaics (PV) experience has been very valuable to me in understanding of diodes, minority carrier dynamics, surface passivation, and photon management. My microelectronics experience is helping me in developing NiSi and tunnel contacts for PV. Microelectronics is becoming extremely advanced and very challenging for a single researcher or for a single organization to explore. Collaborations among institutions are necessary and extremely productive. Technologies evolve, some result in large scales, some address niche applications, and some extinct. Having said that, we learn from our experiences and advance knowledge.

Looking towards the future, microelectronics research needs to focus on requirements for data intensive and edge computing. These will include new materials for logic, memory and storage, heterogeneous integration, new methods of placement and analysis. In addition, new computing architectures and algorithms have to be explored.

Acknowledgements So as I reflect, I believe that it is impossible to realize your dreams and ambitions without others in your corner. Be they inspirational parents, family, dedicated teachers, or encouraging friends. They all stand behind the accomplishments of any individual. I truly believe that it is why I ultimately stayed in education for my career—to keep passing on guidance, support, and learning to future generations. I express my sincere gratitude to my family.

I gratefully acknowledge my PhD advisors Dr. G.C. Jain and Dr. B.K. Das, Postdoc mentor Professor Paul Holloway and former Department Heads—Professor David V. Kerns and Professor Lynn Fuller for their visions and immense support. Special thanks to my colleagues and technical staff of RIT Semiconductor and Microsystems Laboratory. I am very proud of my students and highly value their contributions in the research work conducted in my group. I overwhelmingly thank my collaborators for their wealth of knowledge and support. This work could not have been done without the funding support from agencies—National Science Foundation, Semiconductor Research Corporation, Department of Energy, and from the industry. I am especially thankful to my son, Jivan for keeping *Science Mom* in good humor with love.

Chapter 5
Smart Skin: Multifunctional Flexible Sensor Arrays

Zeynep Çelik-Butler

5.1 Curiosity

5.1.1 Childhood

I was born in Turkey to a medical family, rather late in their lives. As I single child, I received a lot of attention and care from my parents. When told I can be whatever I want to be, my two immediate choices were a belly dancer and a "star traveler"! Curiosity about stars and galaxies led me to devour whatever book I found on the subject. But when it came to choose a career, I was expected to follow my parents' path to medicine, which I had no interest in. I wanted to go into high-energy physics. In an inexplicable compromise between medicine, cosmology, and high-energy physics, it was agreed that I would do a double major in electrical engineering and physics (Fig. 5.1).

I graduated with both degrees from Boğazici University, Turkey in 1982. By now, I had developed interest in solid-state electronics/condensed matter.

5.1.2 Graduate School

Again, curious about what other opportunities lie outside of my borders, I applied to graduate programs in the USA. My intention was to obtain a Masters degree and return to my country. My adviser at University of Rochester, Thomas Hsiang, however, kept throwing scholarships and fellowships at me: IBM Predoctoral Fellowship, Eastman Kodak Fellowship, and others. Before I knew it, I had finished my PhD

Z. Çelik-Butler (✉)
Department of Electrical Engineering, University of Texas at Arlington, Arlington, TX, USA
e-mail: zbutler@uta.edu; http://www-ee.uta.edu/zbutler/

A. C. Parker, L. Lunardi (eds.), *Women in Microelectronics*, Women in Engineering and Science, https://doi.org/10.1007/978-3-030-46377-9_5

Fig. 5.1 Zeynep at age 4. Photo by Seref Celik

with a dissertation on 1/f noise in MOSFETS, and acquired a bonus: a husband, Donald Butler, who worked in the same field as I did.

Although well-known now, the origin of fluctuations leading to low-frequency noise in MOSFETs was undiscovered in early 1980s, and constituted one of the biggest barriers in MOSFET reliability. My work in this field eventually earned my Fellowship in IEEE.

5.1.3 Two-Body Problem

Finding employment in my field with my spouse in a company proved to be almost impossible. Some of the well-known companies at that time in semiconductor industry out-right told us that they will not hire a married couple in a professional position even in different departments. Academia did not have the same issues. We were hired as assistant professors of electrical engineering by Southern Methodist University, Dallas, Texas.

I continued to work on 1/f noise as an assistant and associate professor at SMU. I got involved in Texas Instrument's endeavors of integrating HgCdTe infrared (IR) radiation detectors with read-out circuitry made of the same material. This introduced me into HgCdTe MISFETs, which turned out to be notoriously bad devices. Although the idea was eventually abandoned, my interest in IR sensors was piqued. I started working on uncooled IR sensors in 1990s. After brainstorming with Don

Butler and conversing with experts in DARPA, we started developing the semiconducting phase of YBaCuO as an uncooled bolometer. Soon it was apparent, conventional rigid substrates did not fulfill the needs of all the applications. We started developing flexible sensors.

5.2 Applications and Need for Flexibility

Why would we bother to make flexible, bendable, and even stretchable sensors? Because many current applications demand it as well as potential future uses. Specifically:

- *Wearable sensors*: sensitive gloves for space and industrial applications where the temperature, touch, or flow can be remotely sensed and translated inside the glove for the hands to "feel." This would enable the person wearing the glove to get a feel of the objects and grip them without any bodily harm.
- *Structural health monitoring*: to provide sensor arrays that can be attached to non-planar surfaces such as bridges, aircraft wings, and other vehicles to provide distributed multi-sensory sensing as a step towards real-time structural health monitoring.
- *Wearable physiological monitoring systems*: integrated into a wristwatch, wrist strap, or woven into clothing for diabetics, soldiers on the field, astronauts, or employees working in harsh environments.
- *Minimally invasive surgery (MIS)*: surgical procedures are performed on organs using very small incisions and surgical instruments. If these surgical instruments are equipped with compliant sensors taped on to their outer surface, the surgeon can gain the ability to "feel" the organs, similar to the case of large incision surgeries.
- *Entertainment*: an entrainment suit for full immersion into virtual reality games.
- *Artificial skin*: a flexible, sensing skin for robots. Currently, robotic instrumentation has to function in a structured environment, where everything is limited, to prevent damage to the robot or to the environment by the robot. Another option is to have the robot semi-attended, which defeats the purpose of having a robotic helper. This is necessitated by the lack of distributed and complete sensing capability of robots. If such compliant sensor systems are developed, robotic instrumentation can be made to function self-autonomously on the factory floor, homes, or used for reconnaissance in warfare, security, and defense applications.
- *Sensitive prosthetic devices* such as an artificial leg with a sensitive skin. The sense of temperature, pressure, and flow can be made to feel on another part of the body, or perhaps in the far future, the output signals can be transmitted to the brain for direct sensing. This would provide a sense of feel to prosthetic devices that would help avoid damage and injury.

- *Fingerprint sensors for biometrics*: to measure the characteristic of a finger or hand provided by the unique thermal signature of the blood vessels and groves.
- *Smart tags and labels.*

Most microelectronics utilize a silicon substrate ("wafer") to build the devices on. Then, the die where the devices reside is diced, bonded to electronic packages, and sealed. Realizing fully conformal, bendable, flexible sensors arrays required a completely rethinking for wafer-package system. Borrowing from flexible electronics, which mostly used organic substrates, we started using commercially available polyimide sheets, such as Kapton, as substrates. Packaging was achieved with another polyimide sheet. As shown in Fig. 5.2, if a structure thus packaged in between a flexible substrate and superstrate is bent, the top surface—encapsulation—is under compression while the bottom surface—substrate—is under tension. Therefore, with careful modeling and mechanical simulation, the sensors can be placed on a plane of no strain. This ensures survival of the sensors even when subjected to bending and folding at low radii of curvature.

The minimal requirements that need to be met by sensor packaging are (1) to expose the sensor to the ambient to interact with the environment "transparently," (2) to protect the sensitive electronics from the environment, and (3) to provide vacuum sealing as needed. In addition to these minimal requirements, recent defense and medical application of sensors added the following requirements for sensor packaging, most of which are not met by conventional packaging techniques.

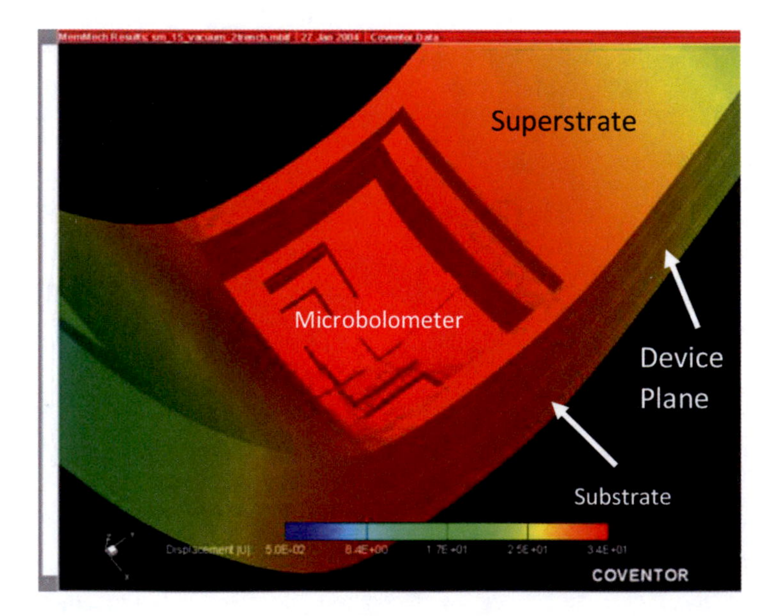

Fig. 5.2 Computer simulation of the sensor structure, flexible substrate and superstrate. For clarity, the superstrate structure has been made invisible. The vertical and horizontal scaling is different

Fig. 5.3 Tactile sensory array rounded around a pen

The microsensors are integrated to give the capability of sensing several stimuli, mimicking a real skin, including but not limited to temperature, touch, and flow. Other requirements are flexibility and ability to stretch and wrinkle without any degradation in its sensing figures of merit. Additional capabilities can be incorporated such as biochemical sensing and decision-making, and energy harvesting. When incorporated with read-out electronics, a complete "Smart Skin" is achieved [1–4].

Use of polyimide sheets as a substrate and superstrate, although flexible and satisfying most of the requirements above, posed a new problem of air gaps and bubbles forming between the fabricated sensors and the polyimide. Another issue was the very restricted thermal budget of the polyimides, severely limiting the processing temperatures for fabricating the sensors. Both problems were simultaneously solved using a carrier wafer on which a liquid polyimide is spin-coated with an appropriate release layer [5] and cured as a substrate, thus avoiding bubbles. Then, all sensor fabrication techniques were redesigned to limit the maximum processing temperature below 325 °C [4, 6]. After fabrication and packaging (superstrate deposition and etching), the silicon wafer is separated from the packaged, finished sensor system.

This enabled development of flexible sensor arrays that can survive being folded down to 5 mm radius of curvature (Fig. 5.3) [7, 8].

5.3 Smart Skin

For the next two decades, I conducted research on sensors on flexible substrates, from radiation to acceleration to tactile to pressure while at the same time working in my original field of noise and fluctuations. This gave me exposure in two distinct fields of electrical engineering and allowed me to obtain funding from several different agencies and companies.

I moved to University of Texas at Arlington in 2002 as a professor of electrical engineering. Due to my managerial experience (3 years) as the Associate Dean of Research and Graduate Studies at SMU, I was chosen as the Director of Nanotechnology Research and Teaching Center at UTA. I served in this capacity from 2003 to 2013, teaching, researching, and directing. I returned to scholarly activities full time in 2013.

As mentioned above, my work on Smart Skin started with flexible microbolometers. From early 1990s to mid 2000s, I worked on infrared (IR) radiation detection.

5.3.1 Flexible Microbolometers

Uncooled microbolometers are utilized for infrared (IR) radiation detection and imaging. IR radiation, coupling with the thermal energy of the atoms (phonons) changes the temperature of the detector, which, in turn, causes a change in the sensor resistance. Our work started with microbolometers on rigid Si substrates [9–12]. This was followed by microbolometers on flexible substrates [13–16]. Microbolometers are one type of thermal detectors, which change their electrical resistance with temperature in response to the radiant energy flux. Another example for a thermal detector is pyroelectric that changes its spontaneous electrical polarization and thereby its surface charge with temperature. In addition, thermopile detectors, also classified as thermal, change their thermoelectric potential with temperature. For a thermal detector, the sensitive element is referred to as the thermometer. The thermometer is typically thermally isolated to improve the responsivity by suspending it above the substrate using micromachining techniques. Figure 5.4 shows the computer model of the developed flexible microbolometer, self-packaged in a vacuum cavity with an IR-radiation-transparent encapsulation, and a resonant cavity, which both enhances the absorption of the detector and thermally isolates the detector from the substrate (heat-sink) for maximum responsivity.

The fabricated flexible detectors exhibited an excellent response to a point IR radiation source as shown in the example (Fig. 5.5).

5.3.2 Flexible Temperature Sensors

Temperature is an important parameter for health monitoring of structures such as bridges, roads, and aircraft, most of which possess non-planar and curved surfaces on which the sensors need to be placed in intimate contact with. Towards these applications, our research group has developed amorphous silicon temperature

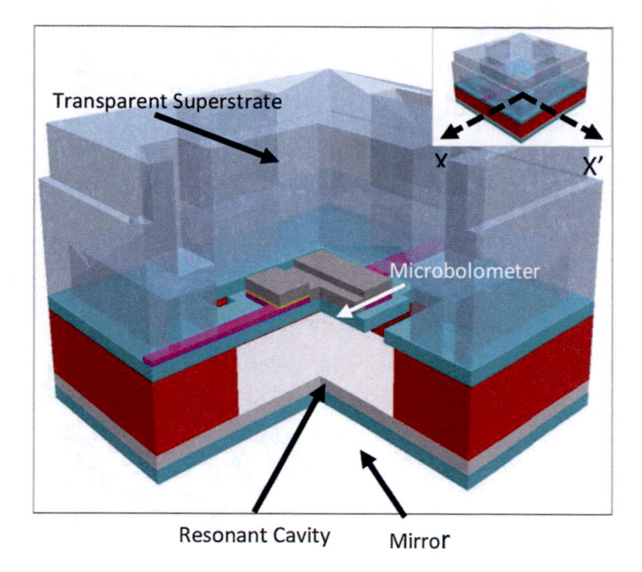

Fig. 5.4 Rendition of a computer-generated model of a microbolometer encapsulated in a vacuum cavity (inset). The main picture shows the cavity sectioned at XX' to show internal details. Substrate is not shown [15]

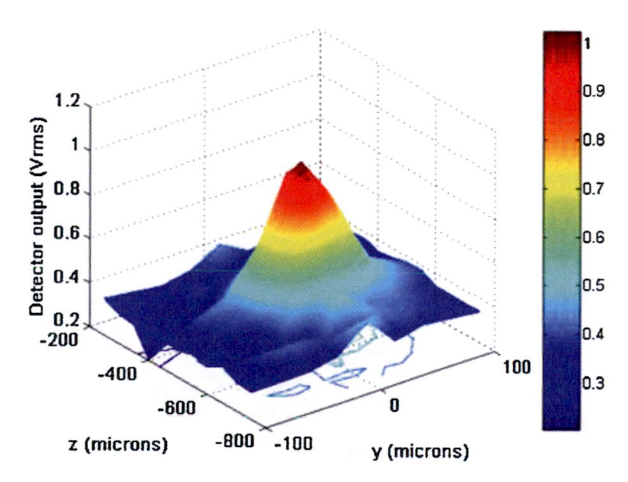

Fig. 5.5 Response of the flexible microbolometer to a point IR source. Detector dimensions were $65 \times 65\ \mu m^2$ [15]

sensor arrays embedded in a flexible substrate [17]. Amorphous silicon was chosen due to its high resistance change with temperature, relatively low noise, the ease of deposition, and compatibility with CMOS technology. The sensor was self-packaged where the flexible polyimide package was grown during the fabrication process and could be directly integrated into flexible circuit boards. The finished sensor array is shown in Fig. 5.6, with the computer generated schematics and the SEM and optical photos in Fig. 5.7.

The obtained sensitivity defined the percentage resistance change per change in temperature was ~2.9%/K, well beyond our initial expectations.

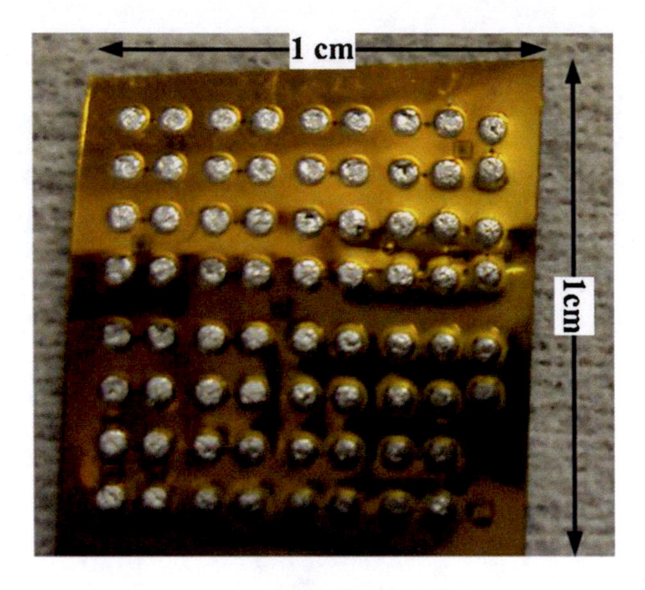

Fig. 5.6 A complete die of temperature sensors in a flexible substrate, containing an array of 35 temperature sensors. The die was removed from the silicon carrier wafer [17]

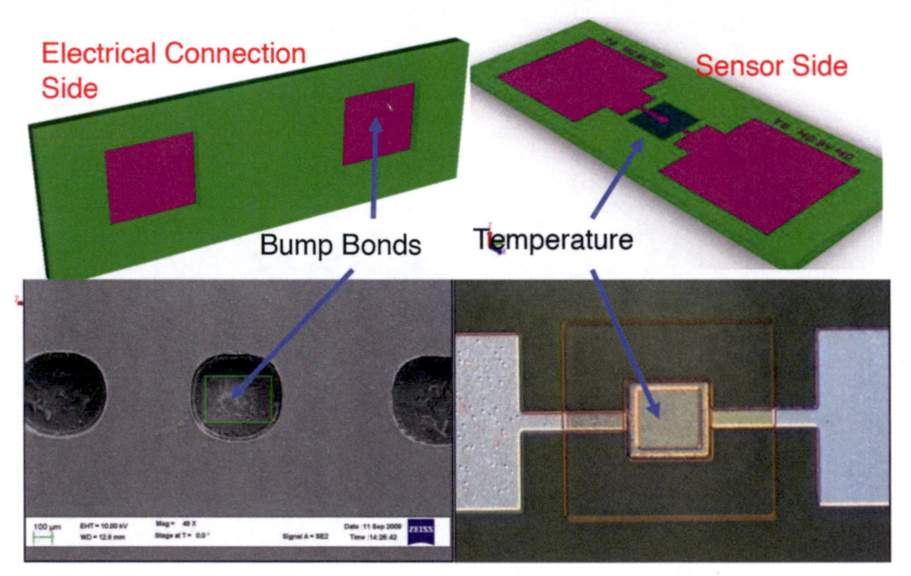

Fig. 5.7 Top: The schematic of the flexible sensor embedded in a polyimide. The sensor side is coated with gold (not shown) for high thermal conductance. Bottom left: The scanning electron microscope of the electrical connection side where bump bonds are to be made. Bottom right: Microscope photo of the thermal sensor and the biasing connections [17]

5.3.3 *Flexible Tactile, Force, and Pressure Sensors*

Perhaps our group in developing flexible force, tactile, and pressure sensors achieved the biggest success. The definition of force is straightforward. Force sensors can be utilized for tactile sensing when they are miniaturized and put into an array form such that the sense of touch is emulated. Pressure sensors, on the other hand, measure force per unit area, but more specifically that of a material in gaseous form. For the sake of organization, we will describe force/tactile sensors together. Then, we will cover the absolute pressure sensors, where the pressure of the gaseous ambient is determined.

As with microbolometers and temperature sensors, our work on force/pressure/tactile sensors started on conventional Si substrates that are rigid. Once low temperature processing techniques were developed [6, 18], we transferred the technology to polyimide-based flexible sensors, using the same wafer-carrier method described in Sect. 5.2.

Initial flexible force/tactile sensor development in our group focused on the well-known Wheatstone configuration, where the resistance of one or two branches of the bridge is varied in response to force, thus resulting in an output voltage change. Different materials such as polysilicon and nickel-chromium were tried as the force sensor piezoresistive material with great success [7, 19].

Our next generation of flexible tactile sensors employed zinc oxide (ZnO) nanorods for their superior spatial resolution, high sensitivity, and self-powering capability [20]. An array of ZnO nanorods was grown using a template-assisted, low-temperature growth process such that the ZnO nanorod crystal c-plane was parallel to the sensor substrate. As a result, the tactile pressure was applied axially to the nanorods, which, in turn, generated a voltage signal with amplitude proportional to the magnitude of the applied pressure. This voltage signal was extracted by means of integrated Ti/Au electrodes deposited at the two ends of the nanorods. The nanorod array and the electrodes were embedded in layers of flexible polyimide (Fig. 5.8), which provided mechanical support, moisture, and dust resistance and

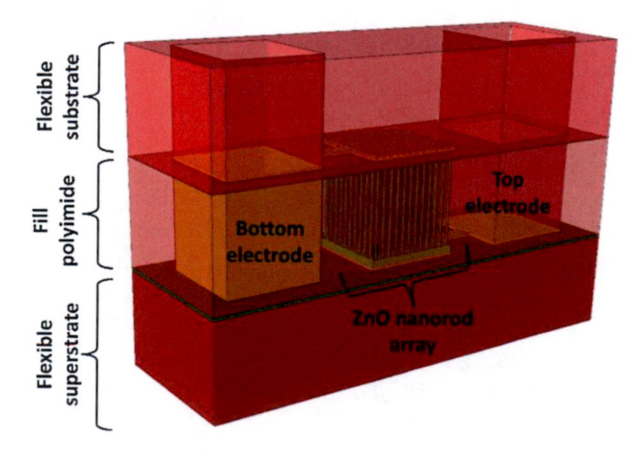

Fig. 5.8 Solid model of tactile sensor structure [20]

Fig. 5.9 Left: The finished robotic skin consisting of array of tactile sensors on a flexible platform. Middle: An addressable tactile sensor array. Right: Close-up of a single tactile sensor made of vertically grown ZnO nanorods acting as self-powered detectors [20]

allowed the sensor array to be draped conformably over non-planar surfaces. A robotic tactile sensor skin thus fabricated is shown in Fig. 5.9 [20].

The need for measurement of absolute pressure requires the sensor to be packaged and sealed so that the structural integrity can be maintained. For flexible absolute pressure sensors, this poses more of a challenge compared to their rigid counterparts, which can utilize conventional hermetic sealing. Typically, a reference pressure is needed for calibration, preferably integrated into the sensor. Our approach to solve both challenges was to place the sensors on a membrane over a sealed cavity with a known pressure. Then, we measured the deflection of that membrane to calibrate the external pressure with the amount of membrane deflection due to any pressure difference between the known and unknown pressures below and above the membrane [21]. Figure 5.10 illustrates flexible, bendable piezoresistive, absolute pressure sensors. Piezoresistors change their resistance in response to force. Nichrome (Ni—80%/Cr—20%) thin-film piezoresistors were arranged on a micromachined Al_2O_3 membrane on top of a vacuum-sealed cavity. The average sensitivity was measured to be 1.25 μV/kPa with an average noise equivalent pressure of 7.44 kPa in the 1/f-noise limit in the 1–10 Hz frequency range and 10 Pa in the Johnson noise limit with a 1 Hz bandwidth [21].

5.3.4 *Flexible Accelerometers*

This was a sensor that we did not think of originally, because it was secondary to other "senses" in our priorities. However, our contacts with funding agencies changed our outlook. It turns out applications in robotics, prosthetics, and other medical fields require functioning flexible micro-accelerometers, encapsulated in vacuum packages, which allow bendability without compromising in performance.

Most accelerometers are mass-spring-damper systems, where the acceleration causes the proof-mass to move, which can be measured and calibrated to the amount of acceleration. Unlike the aforementioned sensors, our group directly moved to

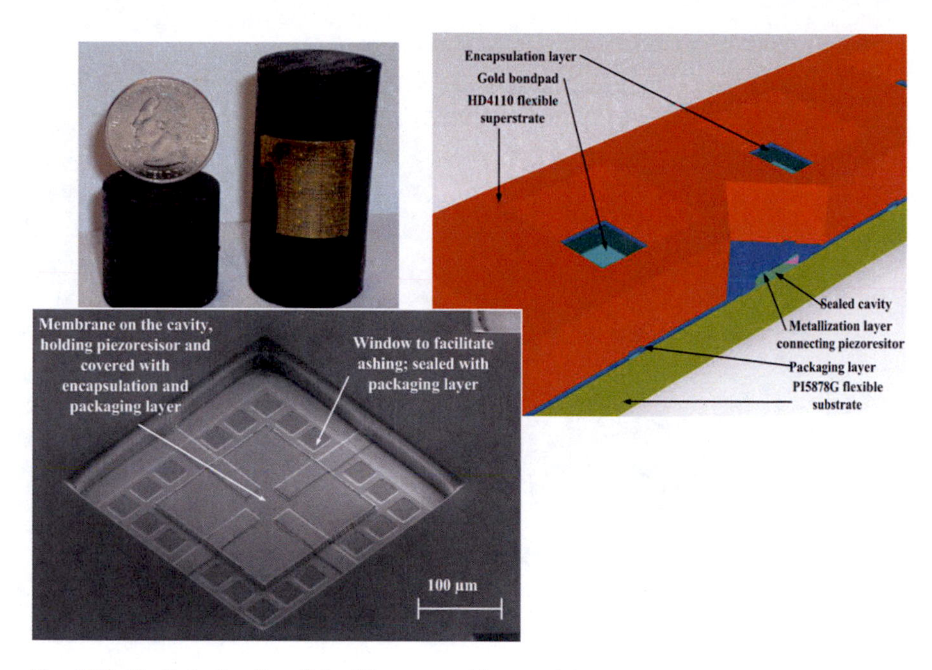

Fig. 5.10 Top left: 2 × 2 cm^2 flexible sensor skin containing 75 pressure sensors. Above: The cross-section of one absolute pressure sensor. Left: The SEM photo of a completed pressure sensor showing the top surface of the membrane where the piezoresistors measure the membrane deflection [21]

developing flexible acceleration sensors without first trying out the rigid ones, since this technology was mature at the time. First the devices were built on polyimide substrates and assessed without packaging [22, 23]. However, without any encapsulation or packaging, the results were limited to laboratory testing but not real-life applications. Then, a flexible polyimide substrate was bonded to a polyimide superstrate. Cavities patterned by Kapton sheet were aligned with the device on a carrier wafer. This carrier wafer was removed upon completion of the packaging and could be re-used, thus lowering the cost [24]. The measured sensitivities of the accelerometers vary between 157 fF/g and 195 fF/g on a flat surface and between 139 fF/g and 174 fF/g when curved down to 2 cm radius of curvature. Here, g is the gravitational acceleration. fF, femtofarad, is the unit of capacitance change. The encapsulation process was found to improve the accelerometer response. Thus packaged and encapsulated flexible accelerometers are depicted in Fig. 5.11.

After completing the encapsulation, the accelerometers were characterized in the same way the open accelerometers were measured [22], on flat and curved surfaces with a radius of curvature (ROC) down to 2 cm. Compatibility of mounting the devices on robotic or prosthetic fingertips was verified. The radius of the robotic fingertips of the index finger and thumb is 1.0 and 3.5 cm, respectively [24].

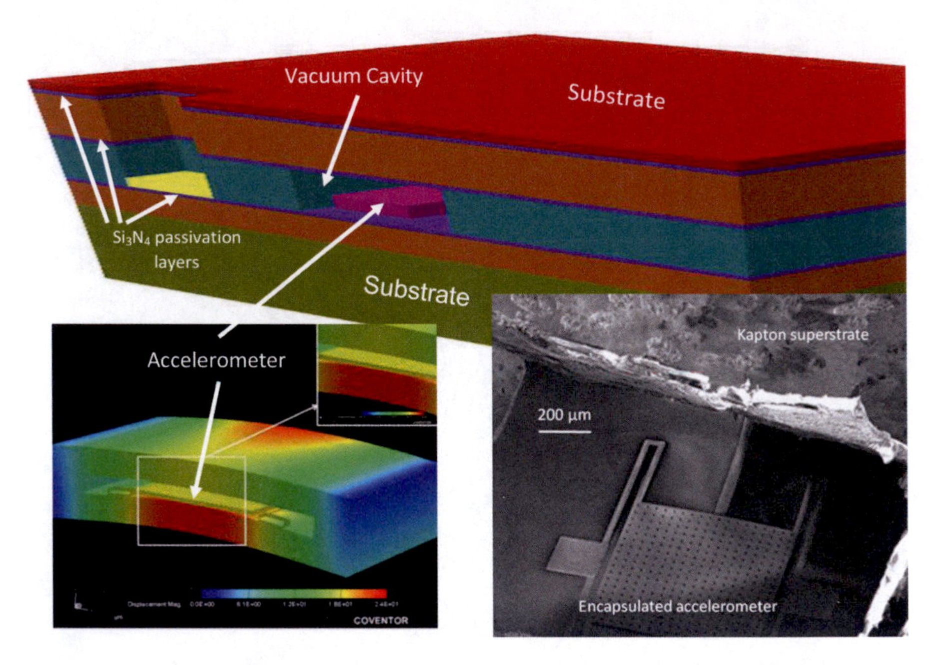

Fig. 5.11 Top: 3D computer rendition of the encapsulated flexible accelerometer. The vertical axis is exaggerated for clarity. Bottom left: Stress simulation of 960 × 960 μm^2 accelerometer when bent to a 2.0 cm ROC. The vertical axis is exaggerated four times. Bottom right: The top encapsulation (superstrate) is lifted off manually to show integrity of the underlying accelerometer after being bent to 2.0 cm ROC [24]

5.4 Summary and Conclusions

I feel privileged to have taken part at the dawn of such an impactful field in sensor electronics.

There is no question that the future of microelectronics is flexible, bendable, and even stretchable, although the last being the furthest achievable. As with general electronics, micro- and nano-sensor systems also require conformity to the platform on which they are placed if their presence is to be unobtrusive and ubiquitous. The accelerated development of Internet-of-Things, together with the aging population ever so dependent on wearable monitoring systems and, in the near future, on robotics, created the need for all-in-one hybrid, flexible integrated circuits and sensors, leading to the development of revolutionary materials and fabrication techniques, some of which have been described above.

The future is flexible.

Acknowledgments I would like to express my appreciation to Dr. Donald Butler. None of this work, of course, would have been possible without my students and post-docs, namely Mouniddin Ahmed, Murali Chitteboyina, Shadi Dayeh, Erkin Gonenli, John Gray, Rohit Kilaru, Yi Li, Sohel

Mahmood, Aamer Mahmoud, Gaviraj Nadvi, Suraj Patil, Vinnayak Shamanna, Christine Travers, Alp Yaradanakul, and Ali Yildiz.

References

1. Çelik-Butler Z, Butler DP (2011) Smart skin: multifunctional sensor arrays on flexible substrates. In: Nalwa HS (ed) Encyclopedia of nanoscience and nanotechnology. American Scientific Publishers. ISBN: 1-58883-188-4
2. Çelik-Butler Z, Butler DP (2012) MEMS on flexible substrates. In: Bhushan B (ed) Encyclopedia of nanotechnology. Springer, Dordrecht. ISBN: 978-90-481-9750-7
3. Butler DP, Çelik-Butler Z (2014) Conformal MEMS sensors. In: Yallup K, Iniewski K (eds) Technologies for smart sensors and sensor fusion. CRC Press, Taylor Francis LLC, Boca Raton, FL. ISBN: 978-1-4665-9550-7
4. Çelik-Butler Z, Dahiya R, Quevedo-Lopez M, Xu Y, Wagner S (2013) Special issue on flexible sensors and sensing systems. IEEE Sensors J 13:3854–3856
5. Ahmed M, Butler DP (2013) Flexible substrate and release layer for flexible MEMS devices. J Vac Sci Technol B 31(5):050602
6. Patil SK, Çelik-Butler Z, Butler DP (2010) Piezoresistive polysilicon film obtained by low-temperature aluminum induced crystallization. Thin Solid Films 519:479–486
7. Nadvi GS, Butler DP, Çelik-Butler Z, Gönenli İE (2012) Micromachined force sensors using thin film nickel chromium piezoresistors. J Micromech Microeng 22:065002. (10pages)
8. Kilaru R, Çelik-Butler Z, Butler DP, Gönenli İE (2013) NiCr MEMS tactile sensors embedded in polyimide, towards a smart skin. J Microelectromech Syst 22:349–355
9. Jahanzeb A, Travers CM, Çelik-Butler Z, Butler DP, Tan SG (1997) A semiconductor YBaCuO microbolometer for room temperature IR imaging. IEEE Trans Electron Devices 44(10):1795–1801
10. Gray JE, Çelik-Butler Z, Butler DP (1999) MgO sacrificial layer for micromachining uncooled Y-B-Cu-O IR microbolometers on Si3N4 bridges. J Microelectromech Syst 8(2):192–199
11. Almasri M, Butler DP, Çelik-Butler Z (2001) Self-supporting uncooled infrared microbolometers with low-thermal mass. J Microelectromech Syst 10(3):469–476
12. Almasri M, Çelik-Butler Z, Butler DP, Yaradanakul A, Yildiz A (2002) Uncooled multi-Mirror broadband infrared microbolometers. IEEE/ASME J Microelectromech Syst 11:528–535
13. Yaradanakul A, Butler DP, Çelik-Butler Z (2002) Uncooled infrared microbolometers on a flexible substrate. IEEE Trans Electron Devices 49:930–933
14. Dayeh SA, Butler DP, Çelik-Butler Z (2005) Micromachined infrared bolometers on flexible polyimide substrates. Sensors Actuators A 118:49–56
15. Mahmoud A, Çelik-Butler Z, Butler D (2006) Micromachined bolometers on polyimide. Sensors Actuators A 132:452–459
16. Mahmoud A, Butler DP, Çelik-Butler Z (2004) Flexible microbolometers promise smart fabrics with imbedded sensors. Laser Focus World 99–103
17. Ahmed M, Chitteboyina M, Butler DP, Çelik-Butler Z (2012) Temperature sensor in a flexible substrate. IEEE Sensors J 12:864–869
18. Patil SK, Çelik-Butler Z, Butler DP (2010) Nanocrystalline Piezoresistive polysilicon film by Aluminum induced crystallization for pressure sensing applications. IEEE Trans Nanotechnol 9:640–646
19. Kilaru R, Çelik-Butler Z, Butler DP, Gönenli IE (2013) NiCr MEMS tactile sensors embedded in polyimide towards smart skin. IEEE/ASME J Microelectromech Syst 22:349–355
20. Nabar BP, Çelik-Butler Z, Butler DP (2015) Self-powered tactile pressure sensors using ordered crystalline ZnO nanorods on flexible substrates towards robotic skin and garments. IEEE Sensors J 15:63–70

21. Ahmed M, Chitteboyina M, Butler DP, Çelik-Butler Z (2015) Flexible, conformal micromachined absolute pressure sensors. IEEE J Microelectromechan Syst 24:1400–1407
22. Gonenli IE, Çelik-Butler Z, Butler DP (2011) Surface micromachined MEMS accelerometers on flexible polyimide substrate. IEEE Sensors J 11:2318–2325
23. Mahmood MS, Çelik-Butler Z, Butler DP (2017) Design, fabrication and characterization of a flexible MEMS accelerometer using multi-level UV-LIGA. Sensors Actuators A Phys 263:530–541
24. Mahmood MS, Çelik-Butler Z (2018) Wafer-level, vacuum-packaged, flexible, bendable accelerometer. IEEE Sensors J 18:4089–4096

Chapter 6
Nanodevices and Applications: My Nonlinear Career Trajectory

Xiuling Li

6.1 Introduction

I am the Donald Biggar Willett Professor in Engineering in the Department of Electrical and Computer Engineering (ECE) at the University of Illinois, Urbana-Champaign (UIUC). I was elevated to be a Fellow of the IEEE, the American Physics Society (APS), and the Optical Society (OSA) in 2017, 2018, and 2019, respectively. My other honors and awards include NSF CAREER award, DARPA Young Faculty Award, and ONR Young Investigator Award. I have published > 150 journal papers, delivered > 120 invited lectures, and hold > 20 patents. My students and postdoc have gone on to become key players in industry including semiconductor, communication, and defense sectors, and faculty members in universities worldwide. I currently serve as the Interim Director of Nick Holonyak Jr. Micro and Nanotechnology Laboratory, the Chair of my department's Promotion and Tenure committee; I am a Deputy Editor for Applied Physics Letters, and Vice President for Finance and Administration at the IEEE Photonics Society. However, I did not start my job as a tenure track Assistant Professor until more than 13 years after my PhD degree, including 6 years working in a startup company without any publications, and all my degrees including my PhD degree are in Chemistry, not ECE.

X. Li (✉)
Micro and Nanotechnology Laboratory, Department of Electrical and Computer Engineering, University of Illinois, Urbana-Champaign, Urbana, IL, USA
e-mail: xiuling@illinois.edu; http://mocvd.ece.illinois.edu/

A. C. Parker, L. Lunardi (eds.), *Women in Microelectronics*, Women in Engineering and Science, https://doi.org/10.1007/978-3-030-46377-9_6

6.2 My Childhood

I grew up in a small town southwest of Beijing in a working class family. My parents never went to college, but my Dad who was very well read and also a highly skilled handyman in the house always encouraged all of us to learn and prioritize school. I am the third daughter of a family with five children. My father coached my eldest sister on her schoolwork, my eldest sister coached my second sister, and I think I just copied their homework and probably everything else. When I was in middle school, my two elder sisters were away in college. With both of my parents working, I was "in charge of" my two younger siblings, who were both in primary schools, for the few hours after school and before my parents returned home—my first leadership experience. We lived in an apartment building so there were always many kids to play with and boss around too. One of my favorite routines on weekends was to go to library, the small one in the chemical plant my father worked at. I do not think there were many science books in that library, but I do recall reading many literature books there. I was a good student and especially excelled at writing in school, so I always thought I would become an author. However, when it came time to declare a major (required during high school application), I chose Science instead of Liberal Arts. Well, I fell for the conventional wisdom because "smart kids" were not supposed to go study literature in those days in China, as the saying was "only when you study Math, Physics, and Chemistry, you can go everywhere in the world." In addition, my Chinese handwriting was horrible, so I decided to aim for a career that deals with numbers not Chinese characters.

I was ecstatic when I got into the best high school in my hometown, knowing where my middle school stood in the ranking. When I started middle school, my family moved from a school district that had the best schools to one that was the closest to where my Dad worked. In that middle school affiliated with the factory, children were spoiled with better facilities but did not have the best resources academically. Most children there did not really work hard since their expectations were to "inherit" their parents' jobs. I went with the flow for more than a year, but subconsciously, I knew that was not what I wanted. I started making weekly bike trips to meet with one of my friends from my primary school, who was attending a good middle school, so that I could get all her homework assignments. That made a huge difference because I would have never seen problems at the high level of difficulties that students in her school were accustomed to. Note that there was no internet back then so the resources were definitely not equally accessible for all, which means that many kids would simply be the products of their local environments. I was also very fortunate to have a teacher, Mr. Linhe Li (not related), who took me to his office one day to show me a newspaper article on study abroad and encouraged me to aim for that.

Unfortunately, the excitement of being in the best high school did not last long. The first assessment exam in that high school gave me the reality check—I almost failed that exam! The way of life in schools back then was ranking, which I believe is still an important system in China nowadays. The students' rankings were based on their scores in every exam and the overall scores in every semester, and the ranking was literally posted in the classroom for everyone to see. Imagine what that could have done to one's confidence! Luckily, within one semester, my ranking went all the way up to number one. I was also the only female on the top eight list, even though that did not register with me until decades later. I must have been oblivious to how girls were perceived, and most of all, I think I was just determined to show myself I could make up the middle school years I goofed off.

6.3 My Education from Two Sides of the Pacific Ocean

Getting into Peking University, one of the most selective colleges in China, was a big milestone. Those were four blissful college years immersed in youthful vibrancy in every aspect and deep intellectual vitality with extraordinary talents from all over the country.

After obtaining my BS degree in Chemistry from Peking University in China, I went to graduate school at the University of California, Los Angeles, again to study Chemistry. Other than not being able to understand Jay Leno's jokes in late night comedy, my transition to the life of a graduate student in the states was quite smooth. In the lab of Prof. Robert Whetten at UCLA, which I had the fortune to join in, I spent a lot of time aligning a pulsed laser beam with a pulsed alkali halide molecular cluster beam in vacuum to measure spectroscopic signatures of the clusters. On the lucky days when the two beams aligned in time and space, I happily stayed in the lab day and night. During some downtimes, I fixed the vacuum pumps, changed gas cylinders, and also socialized with friends. One particular group of friends was four female graduate students, all experimentalists in different groups of Physical Chemistry. Even though there is almost gender parity in Chemistry overall, there were not that many female students in Physical Chemistry. The five of us did a lot of things together, including sharing the ride to conferences, figuring out new tools with longer levers to open the valve of a gas cylinder, celebrating each other's birthdays, semi-periodically making shopping trips to stores like Ross, and occasionally baking cookies for the machine shop staff (to "bribe" them so that our jobs got their attention). This critical group of friends also gave me the best baby shower I could not even have dreamed of when I became pregnant in my last year of graduate school!

After obtaining my PhD degree, I took on a postdoc position in the Department of Chemistry and Chemical Engineering at the California Institute of Technology in the laboratory of Prof. Nate Lewis. In a very short time, I was able to solve a long-standing issue and successfully attached a long chain carbon to a silicon surface without forming oxides at the interface for electron transfer studies. The critical part of the solution was a chemical reaction that I came across while talking to colleagues from another group in the hallway of the Beckman Institute at Caltech. Nate had a big and very active group, and group meetings took place in the evenings. At the first group meeting presentation there, I was warned not to take any questions personally. Sure enough, I was "attacked" with lots of tough questions and the meeting lasted the whole evening, probably past 11 pm. It was intimidating and exciting at the same time; the excitement was what prevailed and the rigorous discussion style in group meetings is what I strive to have to this day.

6.4 Beginning of My Nonlinear Career Path: From Chemistry to Electrical Engineering, and the Gender Factor

Moving to Illinois after a short year at Caltech was not my original intention. I already had an excellent job offer on the west coast at the time. However, with a one-year-old son, I chose to move with my husband to Illinois when he took the faculty position there. As the reluctant trailing spouse, the opportunity to do something different suddenly opened up to me. Instead of staying in chemistry, I decided to take on a second postdoc position in the Department of Electrical and Computer Engineering at Illinois under the direction of Prof. James J. Coleman. I was given the choice of taking on a project that was relatively safe to produce publications or another one that was related to GaN, the "high risk and high payback" one as Jim told me. The choice was a no brainer to me—GaN, the wide bandgap semiconductor that enabled the LED revolution, from epitaxial growth to extensive characterization and device fabrication.

Unbeknownst to me, I was the only female (not counting facility and administration) in the entire clean room where I did my research on GaN. I was very fortunate to have Prof. Coleman who wholeheartedly believed in me. Not an electrical engineer? No problem. Not familiar with MOCVD? No problem. I first attended a GaN workshop in St. Louis where all talks were 5 min only. While lots of people were complaining about the talks being too short, I loved the format and thoroughly enjoyed all the talks, because that was the perfect platform for someone completely new to get a quick grasp of the field. I could not contain my excitement. However, the truth was that so many researchers in the world had already made tremendous progress in the field. How could I, with little background in semiconductors, catch up and get ahead? Somehow, I never doubted that I could. Upon returning to campus, with all the support I had, I dived right into modifying an old reactor used for conventional III-As semiconductors to start growing GaN and soon I became the local expert on GaN MOCVD growth, defects, and device processing and characterization. Years later, when GaN-based LED lamps first became available commercially at a hefty price, I rushed to the store and got some, even though our house was already overstocked with other kinds of bulbs. I just had to. GaN had a special place in my heart and in my mind for reasons beyond scientific ones—it was the first milestone in my career path where I had to get out of my comfort zone to step into a different field and made the technology work to the best I could with limited resources.

6.5 The Discovery of Metal-Assisted Chemical Etching (MacEtch) of Semiconductors: The Ultimate Serendipity

While I worked in the field of III-nitrides as a postdoc, I wrote proposals and went to Washington, DC to talk to funding agencies trying to get support for the program, but with little success. The feedback from panel reviews or program managers included "make her a tenure-track faculty" or "go to Santa Barbara (where a lot of the III-N activities was concentrated)". The fact was I did not want to be a tenure-track faculty then; moving to a different school was not an option, and my young kids were of course conveniently my excuse.

Another opportunity came when the then Head of Chemistry Department, Prof. Paul W. Bohn, kindly offered me a Research Assistant Professor position in Chemistry. The title means nothing more than a super postdoc, but this opportunity took my research to yet another direction—porous semiconductors.

The conventional way to make porous silicon is by anodization, i.e., to apply anodic bias to a silicon wafer typically in hydrogen fluoride (HF) based electrolyte solution. The solution can also include an oxidant such as hydrogen peroxide (H_2O_2). In order to apply the electrical bias, a layer of metal is normally deposited on the back of a piece of conducting silicon wafer. The wafer is then mounted on the bottom of a Teflon cylinder (HF resistant) with an O-ring to seal the electrolyte. A counter electrode Pt is immersed in the electrolyte to form the closed circuit for anodic current generation. It is well known that porous silicon can become direct bandgap and emit red light under UV excitation.

One day, the wafer seemed to have broken during the anodization experiment and the electrolyte (HF + H_2O_2) leaked out of the Teflon beaker to the edge of the wafer (safety hazard!). That piece of wafer happened to have some Au deposited on the front side of one of the corners (recall that metal was supposed to be deposited on the back!), which now was covered by the electrolyte that accidentally leaked out! While I was cleaning up the mess from this disastrous experiment, I decided to check the Si wafer anyway—shining a hand-held UV lamp to see if any red luminescence showed up. Lo and behold—that corner of the wafer, where two mistakes converged (wrong side metal deposition and leaked electrolyte), had the brightest red color I had ever seen! From that point on, I ditched the potentiostat, simply deposited metal (Au or Pt, etc.), and placed the wafer into the electrolyte (HF + H_2O_2) solution under open circuit! That is the serendipity of two mistakes in the lab that day and that is how I discovered metal-assisted chemical etching (MacEtch). Today the paper that documented this discovery, "Metal-assisted chemical etching in HF/H_2O_2 produces porous silicon," Appl. Phys. Lett. 77, 2572 (2000), has been cited over 990 times according to Google Scholar at the time of publication of this article: https://scholar.google.com/scholar?oi=bibs&hl=en&cites=7630633600216444175&as_sdt=5. Nine related patents have been awarded and several more are pending from this discovery and continued development by my research group.

6.6 Life in a Startup Company

It is not easy to hold a non-tenure track position and obtain funding to carry out independent research in a university. After a long postdoc/research assistant professor period of my career, I decided to join a local startup company, EpiWorks Inc. (now II–VI, Inc.). That was not a spur of the moment decision because I always had the itch to work in industry. In fact, I passed on an excellent offer from a company in the bay area before moving to Illinois for family reasons.

Life in a startup was a completely new experience. EpiWorks was in the business of producing epitaxial wafers of GaAs-InGaP based high electron mobility transistors (HEMTs) for preamplifiers in cell phones when it first started. I did not know what statistical process control (SPC) meant for production. I did not know how to interact with customers. I was on call all the time to respond to issues in the fab and release recipes for calibration or production. When our product portfolio expanded to include high-power lasers in a broad range of wavelengths later on, I took on the responsibility of R&D manager. I would take phone calls from colleagues in the fab anytime of the day and any day of the week, including all those time I drove my kids around in evenings and on weekends. My 4-year-old daughter would spontaneously belt out "gas all gas (GaAs/AlGaAs)" and "change the ceiling" into her toy phone! If shipping qualified products to customers on time at the end of each month brought me the sense of satisfaction in completing a task, solving problems was my true calling and real fulfillment at every step of the way in executing customer orders.

6.7 The Start of My Faculty Career and the Innovations Continued with My Research Group

After six years at EpiWorks, I was offered the opportunity to become a tenure-track assistant professor in the department of Electrical and Computer Engineering at UIUC, but I hesitated. I knew I yearned for the intellectual challenge that comes with the freedom of academic research and the far-reaching impact a professor could have in shaping the future workforce. However, I was not a fresh graduate anymore and I had a job that I was content with. I was scared about all the uncertainties, including the possibility of failing to get tenure. When I described this new opportunity at lunch with a friend who was totally outside of my field (a financial analyst), the friend said to me, "you've got to take this job! Your eyes sparkle when you talk about it!" I am proud that, indeed, passion overcome fear and the rest, as the saying goes, was history.

My group's general interests are in the area of semiconductor materials and devices. We focus on developing innovative semiconductor structures and device concepts through both bottom-up and top-down approaches to bring lasting impact to the field of semiconductor electronics, photonics, nanotechnology, quantum technology, and possibly to medicine. Our group has pioneered the development of

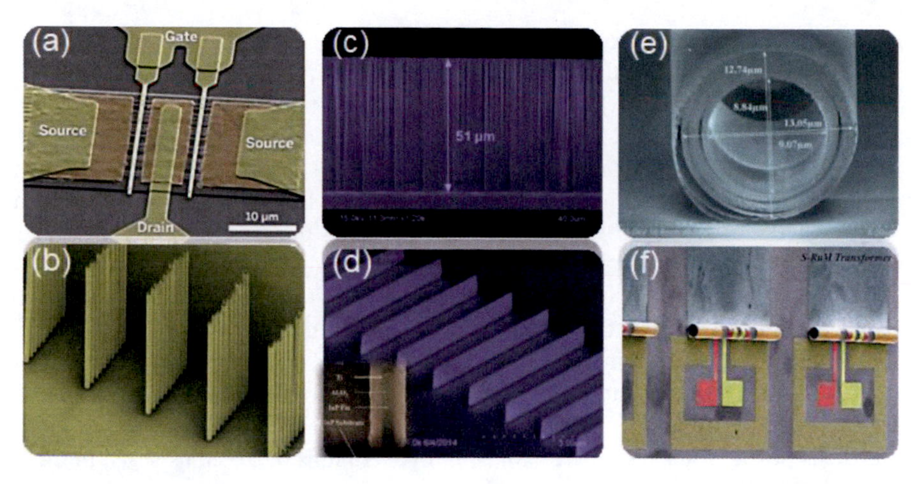

Fig. 6.1 Examples of nanostructures from Xiuling Li's research group: (**a**) a fully fabricated planar GaAs nanowire HEMT structure [1] and (**b**) an array of InAs nanowires on Si by MOCVD (unpublished); (**c**) a high aspect ratio (~100) Si nanowire array [2] and (**d**) an InP nanofin array with a MOSFET cross-section in inset [3], fabricated by the plasma-free metal-assisted chemical etching (MacEtch) method; (**e**) a microtube inductor cross-section view and (**f**) a microtube transformer array [4] for extreme miniaturization of passive electronic components, using the self-rolled-up membrane (S-RuM) platform. Adapted from published work as cited, with permission from ACS, IOP, IEEE, and Springer Nature, respectively

several platform nanotechnologies, as highlighted in Fig. 6.1, including the lateral selective epitaxy of planar nanowire arrays for 3D III–V transistors, the metal-assisted chemical etching (MacEtch) method for damage-free highly anisotropic etching of versatile semiconductor materials and structures, and the self-rolled-up membrane (S-RuM) technology for miniaturization of passive electronic devices. Below are some highlights of my current research portfolios:

6.7.1 Wet Etch, Dry Etch, and Now MacEtch of Semiconductors

From transistors to lasers, etching is an important step in the device fabrication process. Metal-assisted chemical etching (MacEtch) is an unorthodox anisotropic wet etching method that defies the isotropic nature of wet etch through local catalysis effects and enables site-controlled semiconductor nanostructure fabrication with unprecedented aspect ratio (>100:1) [2] and versatility. This innovative etching method has had profound impact on semiconductor fabrication, because of not only the readily achievable extraordinary aspect ratio but also the absence of ion-induced damage. Since our first paper published in 2000 on porous silicon (Si) formation using MacEtch followed by catalyst-site-specific etching, we continue to advance the frontier of this technology from investigating the fundamental mechanism, properties, and applications to extending its applicability to other semiconductors, including germanium (Ge), III-As, III-P, III-N, SiC, and β-Ga_2O_3 [5], as well as heterostructures. Inverse-MacEtch (*i-MacEtch*) allows the formation of atomically smooth sidewalls [3]; magnetic-field guided MacEtch (*h-MacEtch*) enables 3D control of the etching trajectory; self-anchored catalyst MacEtch (*SAC-MacEtch*) promotes the sidewall verticality for large via by using porous catalyst; UV-assisted MacEtch (*UV-MacEtch*) makes plasma-free wide-bandgap semiconductor etch possible; novel catalysts including titanium nitride (TiN) and ruthenium (Ru) ensure CMOS compatibility of the MacEtch process; and the ultimate vapor phase MacEtch (*VP-MacEtch*), while maintaining the damage-free nature, truly taking the technology towards scalability and manufacturability. The simplicity, versatility, manufacturability, and realistic potential of MacEtch to replace and enhance dry etch methods position this technology for future generations of 3D transistors, through-silicon-vias, trench memory, thermoelectric detectors, and photovoltaic devices, as well as biomedical devices.

6.7.2 III–V Nanowire Array for Beyond Si CMOS, RF Electronics, and Quantum Technologies

Nanowires have been considered as a promising architecture for beyond Si CMOS logic and future III–V RF electronics, as well as next generation optoelectronic applications. The challenges have been the controllability and manufacturability. Our discovery of parallel arrays of planar III–V nanowire growth mode opens up a new paradigm of crystal growth: selective lateral epitaxy and consequently in situ lateral junctions. Technologically, in-plane nanowire configuration is perfectly compatible with existing planar processing technology for industry. Chip-scale transistors (HEMTs) with record DC and RF performance are demonstrated [1], with a clear path to reach THz for high-speed applications. Low bandgap nanowires also play a significant role in topological quantum devices by coupling to superconductors [6].

6.7.3 MOCVD Epitaxial Heterogeneous Integration of III–V, Si and Layered van der Waals Sheets

The quintessential challenge of heterogeneous integration of III–V thin films on silicon and other foreign substrates has been the lattice-matching restriction. Owing to the small hetero-interfacial area footprint of a nanowire on a substrate and the free sidewall facets, it has been well demonstrated that semiconductor nanowires are amenable to direct and high quality heteroepitaxial crystal growth of materials with highly mismatched lattice constants and coefficients of thermal expansion. We have achieved site-controlled selective area epitaxy by metalorganic chemical vapor deposition (MOCVD) and characterization of GaAs, InAs, InGaAs, GaP, GaAsP nanowires on GaAs, Si, as well as 2D van der Waals surfaces [7] including graphene, MoS_2, and BN. These unconventional epitaxial growth modes my lab pioneered are positioned to bring translational impact on nanoelectronics and nanophotonics, including III–V gate-all-around transistors and multi-junction tandem solar cells.

6.7.4 Self-Rolled-Up Membrane (S-RuM) Nanotechnology for Extreme Miniaturization and Integration of Passive Electronics, Photonics, and Biology

The overarching physical principle of S-RuM nanotech is strain-driven spontaneous deformation of 2D membranes into 3D architectures. Complex 3D structures enable advanced functionalities that are otherwise out of reach. S-RuM inductors have been demonstrated with a footprint that is 10–100 times smaller than the 2D counterpart and capable of operating at high frequencies (>20 GHz) [8]. With further agressive scaling in the number of coil turns and post-rolling core-filling, we have realized tens of millitesla level magnetic induction [9]. By stacking two S-RuM inductors in-plane or vertically to form transformers, near unity coupling coefficients and unprecedentedly high turn-ratios have been achieved [4]. Through global and local strain engineering, S-RuM filters, transmission lines, antennas with ultra-high frequency (including THz), and bandwidth can all be enabled. S-RuM technology promises to remove the constraints of size, weight, and performance (SWAP) of RFICs and even millimeter wave communications. The unique form factor can also bring translational impact to wearable and flexible IoT devices and other exciting opportunities including guiding and accelerating neuron cell growth using S-RuM for neural regeneration [10].

6.8 Closing Remarks

My career trajectory is not linear and the changes at every step were not always planned: from a Chemistry degree to a postdoc in Electrical Engineering, from a postdoc to a startup company, and from a startup to climbing the tenure ladder from assistant to associate to full professor. I am grateful for all the guidance and opportunities that were provided to me by many amazing mentors and friends I have been so fortunate to encounter. I am proud that I embraced those opportunities, overcame fears, and turned them into tangible outcomes, while doing the balancing act of family and career. I will be more proud if my mentoring to others, especially young female graduate students, can guide and inspire them to pursue their dream careers, no matter how many conformable zones they have to step out of!

References

1. Miao X, Chabak KD, Zhang C, Mohseni PK, Walker DE Jr, Li X (2015) High speed planar GaAs nanowire arrays with fmax > 75 GHz by wafer-scale bottom-up growth. Nano Lett 15(5):2780–2786
2. Balasundaram K, Sadhu JS, Shin JC, Azeredo B, Chanda D, Malik M, Hsu K, Rogers JA, Ferreira P, Sinha S, Li X (2012) Porosity control in metal assisted chemical etching of degenerately doped silicon. Nanotechnology 23:305304
3. Song Y, Mohseni PK, Kim SH, Shin JC, Ishihara T, Adesida I, Li X (2016) Ultra-high aspect ratio InP junctionless FinFETs by a novel wet etching method. IEEE Electron Dev Lett 37(8):970–973
4. Huang W, Zhou J, Froeter P, Walsh K, Liu S, Kraman MD, Li M, Michaels JA, Sievers DJ, Gong S, Li X (2018) Three-dimensional radio frequency transformers based on a self-rolled-up membrane platform. Nat Electron 1:305–313
5. Huang H-C, Kim M, ZhanX, Chabak K, Kim JD, Kvit A, Liu D, Ma Z, Zuo J-M, Li X (2019) High Aspect Ratio β-Ga O Fin Arrays with low-interface charge density by inverse metal-assisted chemical etching. ACS Nano 13(8):8784–8792
6. Finck ADK, Van Harlingen DJ, Mohseni PK, Jung K, Li X (2013) Anomalous modulation of a zero-bias Peak in a hybrid nanowire-superconductor device. Phys Rev Lett 110(12)
7. Mohseni PK, Behnam A, Wood JD, Zhao X, Yu KJ, Wang NC, Rockett A, Rogers JA, Lyding JW, Pop E, Li X (2014) Monolithic III-V nanowire solar cells on graphene via direct van der waals epitaxy. Adv Mater 26(22):3755–3760
8. Yu X, Huang W, Li M, Comberiate TM, Gong S, Schutt-Aine JE, Li X (2015) Ultra-Small, High-Frequency and Substrate-Immune Microtube Inductors Transformed from 2D to 3D. Scientific Reports 5(1)
9. Huang W, Yang Z, Kraman MD, Wang Q, Ou Z, Rojo MM, Yalamarthy AS, Chen V, Lian F, Ni JH, Liu S, Yu H, Sang L, Michaels J, Sievers DJ, Eden JG, Braun PV, Chen Q, Gong S, Senesky DG, Pop E, Li X (2020) Monolithic mtesla-level magnetic induction by self-rolled-up membrane technology. Sci Adv 6(3):eaay4508
10. Froeter P, Huang Y, Cangellaris OV, Huang W, Dent EW, Gillette MU, Williams JC, Li X (2014) Toward intelligent synthetic neural circuits: directing and accelerating neuron cell growth by self-rolled-up silicon nitride microtube array. ACS Nano 8(11):11108–11117

Chapter 7
Gate-All-Around Silicon Nanowire Transistor Technology

Ru Huang, Runsheng Wang, and Ming Li

7.1 Introduction

Even only from the viewpoint of geometric aesthetics, gate-all-around silicon nanowire transistor (GAA SNWT) is undoubtedly the most promising candidate for the ultimately scaled device toward the end of roadmap. Due to the ideal gate controllability, excellent transport properties and feasible device design, GAA SNWT has provided a realistic chance for the CMOS technology to advance into sub-3 nm era or even be applied earlier (http://irds.ieee.org). Not only the great potential in the core circuits of CMOS VLSI, SNWT also provides other technology diversities such as I/O devices, RF/Analog circuit, sensors, and so on.

However, so many attractive advantages for SNWT, there are still a lot of challenges in process fabrication and device understanding, such as controlling the shape of nanowire, variability and reliability, and other parasitic secondary effects. In the following sections, our research results addressing those issues will be presented [1]. In the second section, the basic fabrication and characterization of SNWT will be introduced and then the extensive investigation of quasi-ballistic transport, self-heating effect, and predictive parasitic capacitance modeling will be discussed. In the third section, variability and reliability issues of SNWT, especially LER/LWR, flicker noise, and random telegraph noise will be discussed. In the fourth section, several important application examples including electrostatic discharging protection diode, RF/Analog circuit, and sensor device will be given to demonstrate the potential of SNWT in technology diversity.

R. Huang (✉) · R. Wang · M. Li
Peking University, Beijing, China
e-mail: ruhuang@pku.edu.cn

A. C. Parker, L. Lunardi (eds.), *Women in Microelectronics*, Women in Engineering and Science, https://doi.org/10.1007/978-3-030-46377-9_7

7.2 Fabrication, Characterization, and Modeling of SNWTs

7.2.1 *Fabrication of SNWTs on Bulk Si*

From the integration viewpoint, controlling the shape of nanowire is the most challenging technique for fabricating SNWTs since any deviation from the designed dimension will cause serious variation problem. Therefore, a stable and self-convergence process is required to fabricate the nanowires according to the designed dimensions.

We have proposed and demonstrated a kind of self-limiting oxidation technique to control the size and shape of nanowires [2]. This technique is originated from the retarded oxidation due to the stress-induced reaction rate reduction and high-density-oxide-induced diffusivity reduction. As shown in Fig. 7.1, the silicon fins are firstly patterned and covered by the nitride liners and then the neck region is isotropically wet etched to form the initial nanowire with water-drop-like shape. Long time wet oxidation is then carried to drive the nanowire to transform into an ideal circular shape. Figure 7.2 shows the scan electron microscope images of the initial suspended nanowire and the final cross-section after self-limiting oxidation. It should be noted that this technique is without any epitaxy processes for sacrificial layer growth to also show the cost effectiveness.

One of the keys to control the shape of nanowire accurately by self-limiting oxidation is the modeling capability to predict the transformation of nanowire cross-section. By incorporating the stress and orientation dependent reaction rate and diffusivity into classic Deal–Grove model, we have set up a calibrated simulation flow for the self-limiting oxidation [3]. The corner rounding effect and two-dimensional oxidation process can be simulated accurately. The calibrated model has precisely simulated the intermediate shape deformation during self-limiting oxidation process even down to nanometer scale, as shown in Fig. 7.3.

With the self-limiting oxidation scheme, the 10-nm-diameter SNWTs are fabricated on the bulk silicon substrate [2], as shown in Fig. 7.4. In the fabrication flow, a bottom channel implantation with dose of 1×10^{18} cm^{-2} is carried to avoid the turn-on of parasitic transistor.

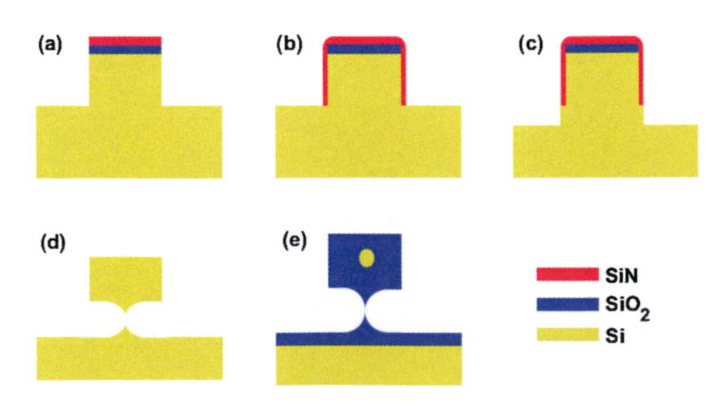

Fig. 7.1 (**a–e**) schematic diagrams of main flow to fabricate silicon nanowires by self-limiting oxidation

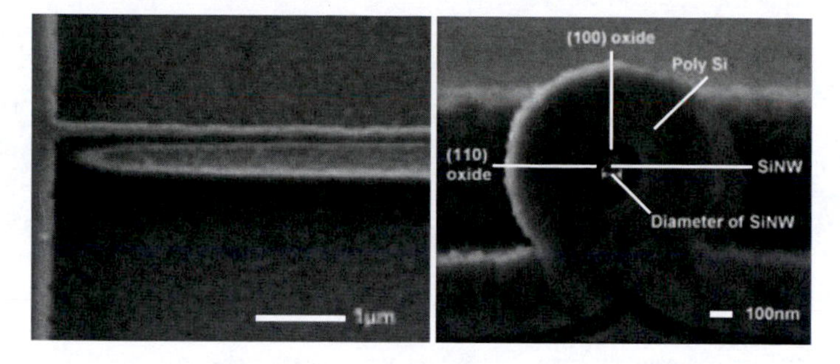

Fig. 7.2 The scan electron microscope images of (**a**) initial nanowire and (**b**) final cross-section after self-limiting oxidation

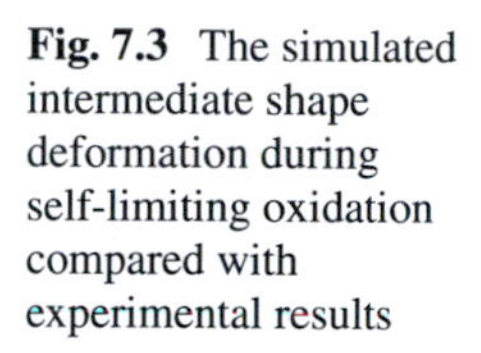

Fig. 7.3 The simulated intermediate shape deformation during self-limiting oxidation compared with experimental results

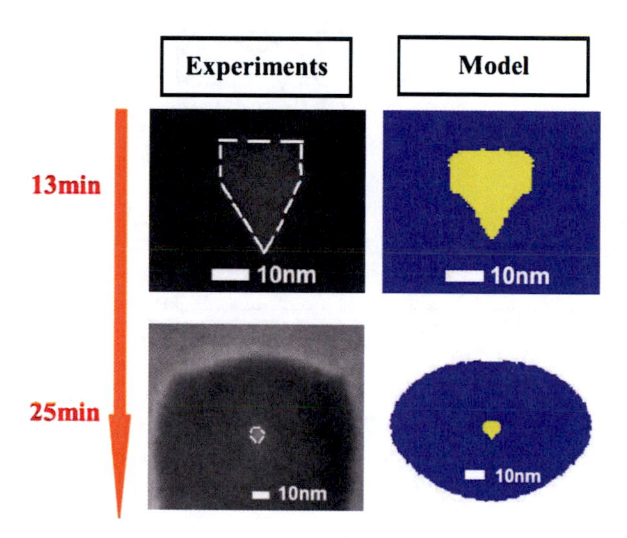

The measured transfer and output characteristics of the fabricated SNWTs are shown in Fig. 7.5. The excellent short channel effect immunity and the highest I_{on}/I_{off} ratio up to 2.6×10^8 have been achieved due to the very small nanowire diameter and perfect nanowire/oxide interface provided by self-limiting oxidation.

Despite 5 nm gate oxide, high driving current of 1.04 mA/μA and low off-state leakage current of 12fA/μm are still obtained with the drain-induced-barrier-lowering of 4 mV/V and subthreshold swing of 74 mV/decade. Figure 7.6 shows the comparison of electron mobility between SNWT and double-gate MOSFET [2]. Due to the near-center transport inside nanowire channel and the improved surface roughness by self-limiting oxidation, the electron mobility of SNWT is significantly higher than that of double-gate MOSFET. The advantage in carrier mobility can provide the potential for SNWT to work in quasi-ballistic transport regime to obtain high transport performance, as will be discussed in the next section.

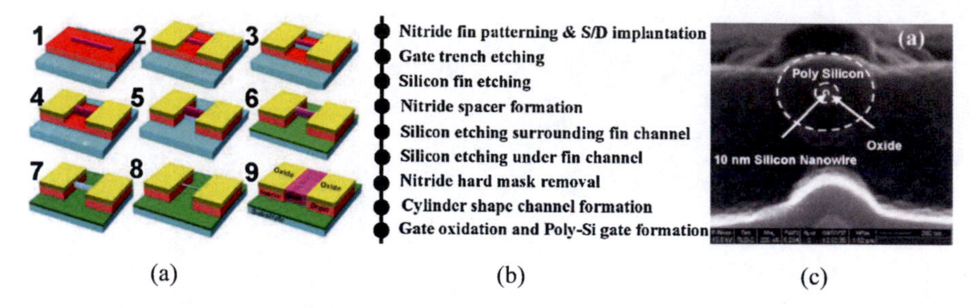

Fig. 7.4 (**a**) The schematic diagrams and (**b**) the key process steps of the SNWTs fabrication on bulk silicon substrate by self-limiting oxidation, and (**c**) the cross-sectional scan electron microscope image of 10-nm-diameter nanowire with 5 nm gate oxide and polysilicon gates

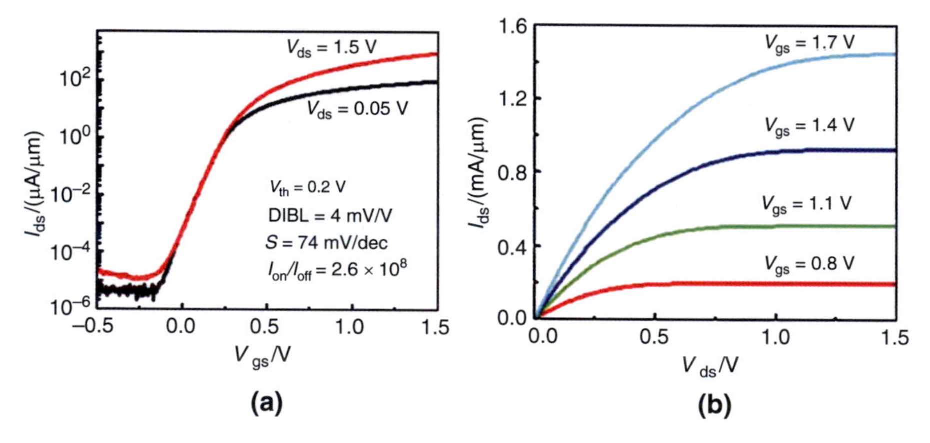

Fig. 7.5 (**a**) the transfer characteristics and (**b**) the output characteristics of SNWTs fabricated on bulk silicon substrate by self-limiting oxidation process

7.2.2 *Quasi-Ballistic Transport in SNWTs*

Quasi-ballistic transport is one of the key features of SNWTs to enable the high performance. According to Lundstrom's channel backscattering theory [4], the saturation current of a nanoscale device can be expressed by the characteristic factor of ballistic efficiency B_{sat} as:

$$I_{D,\text{sat}} = WC_{\text{eff}}v_{inj}\left(\frac{1-r_{\text{sat}}}{1+r_{\text{sat}}}\right)\left(V_G - V_{T,\text{sat}}\right), r_{\text{sat}} = \frac{1}{1+\frac{\lambda_0}{l_0}}, B_{\text{sat}} = \frac{1-r_{\text{sat}}}{1+r_{\text{sat}}} \tag{7.1}$$

where $I_{D,\text{sat}}$ is the drain saturation current, W the channel width, C_{eff} the effective gate capacitance, v_{inj} the carrier injection velocity at the virtual source as shown in Fig. 7.7, V_G and $V_{T,\text{sat}}$ the gate bias and saturation threshold voltage, respectively, and

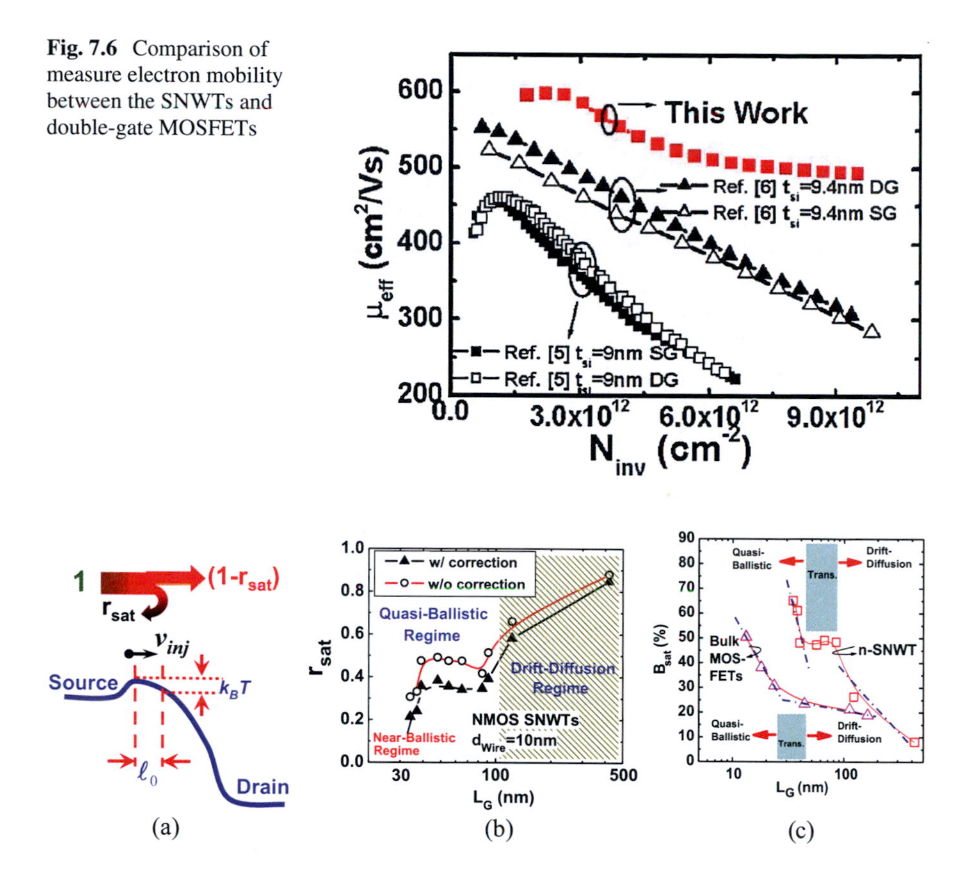

Fig. 7.6 Comparison of measure electron mobility between the SNWTs and double-gate MOSFETs

Fig. 7.7 (**a**) The schematic diagram of Lundstrom's channel backscattering theory. (**b**) The extracted r_{sat} in n-channel SNWTs with and without the temperature dependence of R_S. (**c**) The extracted B_{sat} of the fabricated SNWTs and bulk MOSFETs

r_{sat} the backscattering rate in the channel, λ_0 the mean-free-path length, l_0 the key distance along which the channel potential drops about k_BT.

Based on Lundstrom theory, a kind of temperature dependent I–V technique was developed to extract r_{sat} and B_{sat} [5]. This extraction method makes an assumption that the temperature dependence of device's source resistance R_S is negligible, which is appropriate to the planar devices with channel and source/drain (S/D) regions in similar dimensions. However, for the quasi-1D nanowire structures, which are connected to 3-D S/D regions in an abrupt way, the temperature dependence of R_S cannot be ignored. This is because the quantum contact resistance, which is a part of R_S and originates from the interface between the 3-D S/D regions and quasi-1D quantum-wire of S/D-extension regions, is sensitive to temperature. Thus, we have proposed a modified experimental extraction methodology for SNWTs [6, 7], in which the temperature dependence of source resistance has been included. Figure 7.7b compares the extracted r_{sat} for SNWTs with or without the above correction. It can be explicitly seen that severe overestimation occurs if the temperature dependence of R_S is neglected. Figure 7.7c further plots the extracted B_{sat} of the SNWTs comparing with conventional planar bulk MOSFETs as functions of the gate length [8]. The ballistic efficiency exhibits remarkable increase especially when SNWTs entering the sub-40 nm region, which is earlier than bulk MOSFETs, due to larger low-filed mean-free-path in SNWTs. The extracted carrier injection velocity v_{inj} at the source is almost unchanged due to the thermal injection limit, which indicates the possibility for further performance enhancement with some strain engineering.

Figure 7.8 shows the ballistic limit, the apparent mobility model predictions as well as experimental data of n-channel and p-channel SNWTs [6]. It is clearly seen that GAA SNWTs are approaching the ballistic limit as the gate length shrinking. Due to the combination of 1-D nanowires and gate-all-around structures, the nanoscale GAA NWFET can be one of the promising structures to provide the possibility of near-ballistic MOSFET from top-down approach.

What is more, the emerging of plateau part of B_{sat} in Fig. 7.7 can be explained as the result of competition between diffusion-drift transport and quasi-ballistic transport. According to the modified Shur's model [6], the apparent mobility of SNWT can be expressed as the function of long channel mobility μ_{long}, mean-free-path λ_0, and gate length L_G.

$$\mu = \frac{\mu_{long}}{\frac{2\lambda_0}{L_G} + 1} \tag{7.2}$$

By extracting the apparent mobility with different gate length in Fig. 7.8, the mean-free-path λ_0 can then be calculated to be about 43.3 nm for n-type SNWTs and 45.3 nm for p-type SNWTs. According to the Lundstrom's channel backscattering theory, the device will enter the regime of quasi-ballistic transport as $L_G < \lambda_0$ and the regime of diffusion-drift transport as $L_G > 2\lambda_0$. Between λ_0 and $2\lambda_0$, just right in the

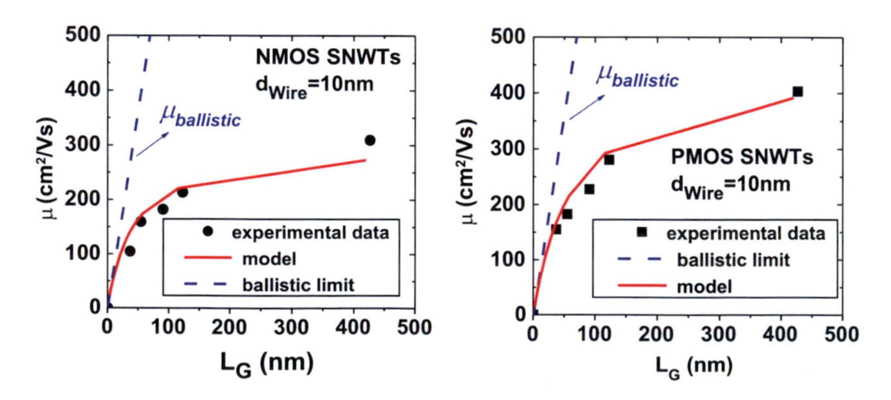

Fig. 7.8 Comparison among the ballistic limit, apparent mobility model predictions and the experimental data of (**a**) n-channel SNWTs and (**b**) p-channel SNWTs

plateau region, two transport mechanisms will compete each other. With the state-of-the-art technology, the modern SNWTs usually work in the quasi-ballistic regime to show performance advantage over the conventional planar MOSFET.

7.2.3 Modeling of Parasitic Capacitances of SNWTs

The extrinsic parasitic capacitance has emerged as a scaling limitation for SNWTs [9]. The 3-D schematic and the 2-D cross-sectional views of a typical GAA SNWT are shown in Fig. 7.9. The cylindrical nanowire is extended to connect the undoped channel and the heavily-doped source/drain to form the source/drain extension (SDE) region. The parasitic gate capacitances of the SNWT are then divided into 4 parts: outer fringe capacitance C_{of}, inner fringe capacitance C_{if}, overlap capacitance C_{ov}, and sidewall capacitance C_{side} [10].

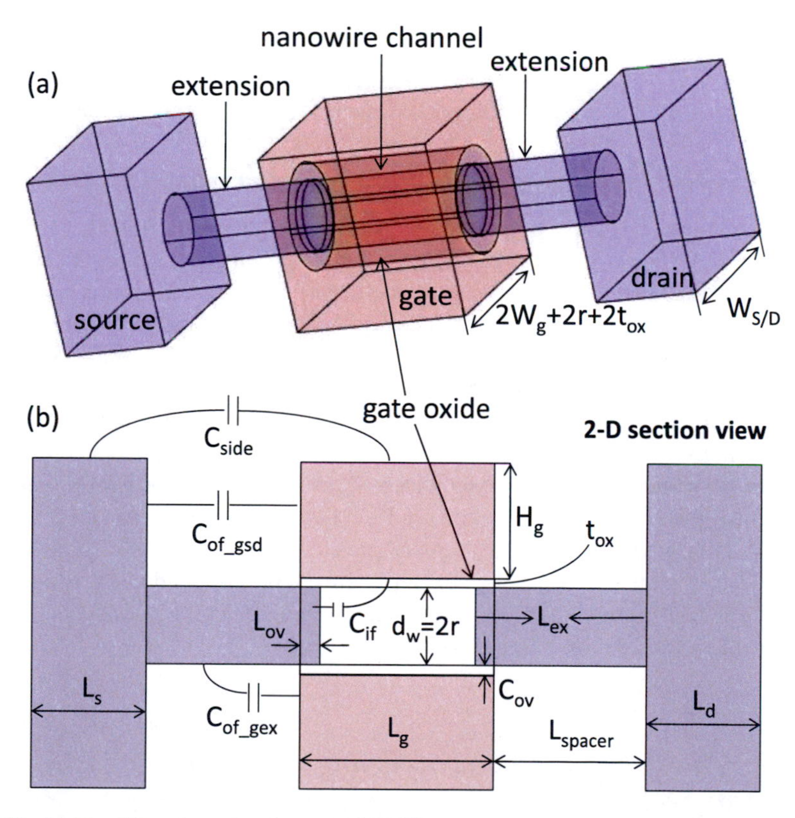

Fig. 7.9 (**a**) The 3-D schematic of a typical SNWT structure and (**b**) the 2-D sectional view of SNWT along the channel direction. H_g represents the height of the gate, and W_g represents the width of the gate along the horizontal direction. The length of the source/drain is denoted by $L_{s/d}$. The parasitic gate capacitances are divided into four parts: C_{of}, C_{side}, C_{ov}, and C_{if}, where C_{of} can be regarded the superposition of C_{of_gsd} and C_{of_gex}

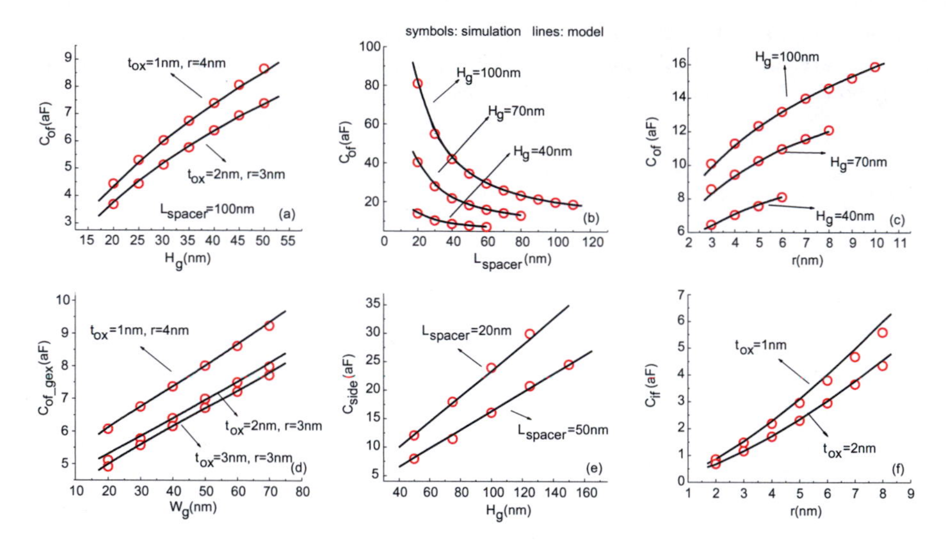

Fig. 7.10 Comparison between the parasitic gate capacitance model (lines) and the 3-D numerical simulation results (symbols) as a function of various device parameters. (**a**) C_{of} as a function of H_g at large L_{spacer}. (**b**) C_{of} as a function of L_{spacer} at different H_g. (**c**) C_{of} as a function of r. (**d**) C_{of_gex} as a function of W_g at different r and t_{ox}. (**e**) C_{side} as a function of H_g at different L_{spacer}. (**f**): C_{if} as a function of r. W_g is assumed to be equal to H_g in (**a–f**) for convenience

A predictive compact model has been proposed for the parasitic gate capacitances of GAA SNWTs [10], which has no fitting parameters. The 3-D capacitance system is calculated by the equivalent transformation and the inversion of Schwarz–Christoffel mapping. The model agrees well with the results of 3-D electrostatic numerical simulations, as shown in Fig. 7.10.

The results also show that the proportion of parasitic gate capacitances in the total capacitance is increased in the gate-all-around architecture due to the ultra-small dimension of the SNWT channel, and thus the proportion of intrinsic capacitance is reduced. Among the capacitances, C_{of} is found to be the largest contributor to the total parasitic gate capacitance in FO1 delay calculation, and C_{side} manifests itself as a non-negligible parasitic capacitance.

The developed capacitance model can be easily incorporated into a compact core model of SNWTs, such as BSIM-CMG, for further device/circuit design optimizations with various device parameters.

7.2.4 Characterization and Modeling of Self-Heating Effect of SNWTs

With MOSFETs scaling into nano-region, increased self-heating effects (SHE) become a critical concern for device performance and reliability degradation, due to the adopted materials with low thermal conductivity (SOI, SiGe, etc.) or/and the

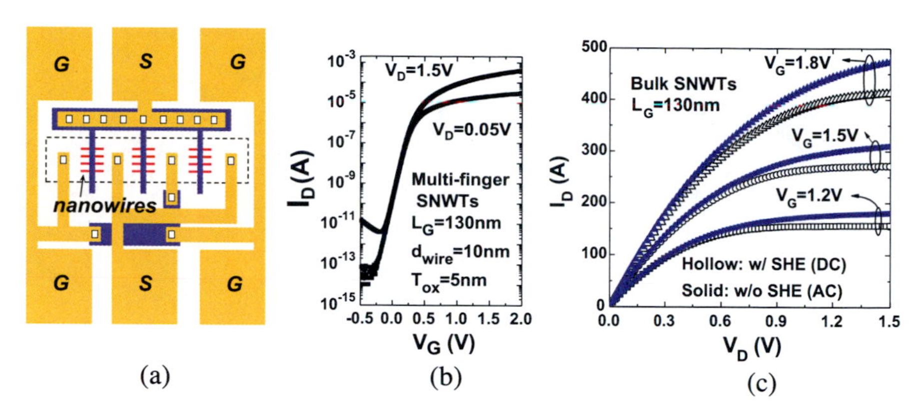

Fig. 7.11 (**a**) The schematic view of multi-finger and multi-wire SNWTs with the G-S-G test structure for high-frequency measurements. Each SNWT has 4 nanowires connected in parallel. (**b**) The typical measured I_D-V_G characteristics of the SNWT with 12 nanowires and 3 gate fingers. (**c**) The measured I_D-V_D with and without self-heating effects in 130 nm SNWTs fabricated on fully bulk silicon substrate

physical confinement of the new device structures (UTB, FinFET, etc.). In GAA SNWTs, the problem may be even worse due to their ultra-small channel and strong-confined structure.

With multi-finger and multi-wire test structure, the impact of self-heating effect is successfully characterized [11], using the AC conductance technique, as shown in Fig. 7.11.

The results indicate that, even if the SNWT is fabricated on bulk silicon substrate, the impact of self-heating effect is comparable or even a little bit worse than SOI devices, as shown in Fig. 7.12 [11]. This is due to the structural nature of SNWTs as follows: (1) The SNWT has small silicon thickness, and its ultra-scaled dimension of nanowire structure results in increased power density, due to the enhanced current drivability and limited silicon volume for heat dissipation in SNWTs. (2) Because of the strongly confined nanowire structure, there are limited modes for heat dissipation in SNWTs. The heat transport in nanowire channel is in quasi-one dimension. Moreover, additional thermal contact resistance exists at the abrupt interface between 1D-nanowire channel and 3D-S/D region, which can modify the phonon modes participating in energy transmission and thus create an additional thermal resistance. While this parasitic contact thermal resistance does not exist in planar devices, in which the channel and S/D region are in similar dimensions without abrupt change. (3) The surrounding gate architecture and increased surface to volume ratio of SNWTs can induce strong phonon-boundary scattering and thus increased boundary thermal resistance, which is worse than UTB SOI, double-gate, or tri-gate structures.

In order to accurately describe the self-heating effects in GAA SNWTs for thermal-aware design optimization, we have proposed an electro-thermal model [12], which is derived based on the equivalent thermal network method, as shown in Fig. 7.13. In this model, the impacts of gate length, nanowire diameter, and

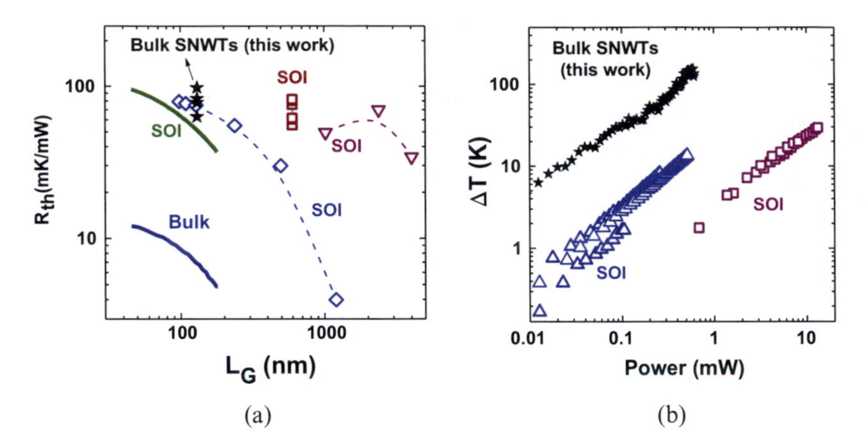

Fig. 7.12 (**a**) The measured R_{TH} of SNWTs, comparing with the results of SOI devices and bulk MOSFETs. (**b**) The increased average channel temperature in SNWTs as a functional of power, compared with SOI devices

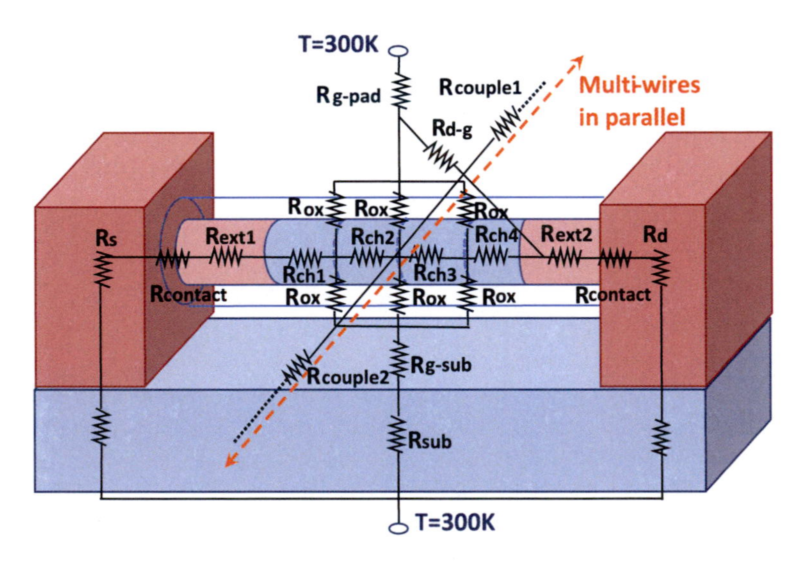

Fig. 7.13 The schematic of a GAA SNWT on bulk substrate with equivalent thermal resistances

surface roughness on the nanowire channel thermal conductivity as well as the influence of unique GAA structure features on the heat dissipation are taken into account. It agrees well with the experimental results of SNWTs, as shown in Fig. 7.14. This model can be further applied to the thermal-aware design of SNWT-based circuits.

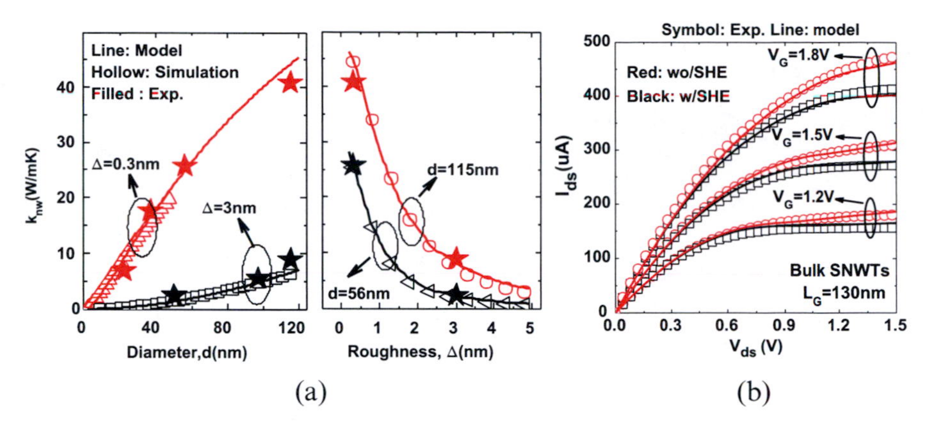

Fig. 7.14 (**a**) Comparison of silicon nanowire thermal conductivity between the model and simulation as well as the experimental results. (**b**) The electro-thermal modeled drain currents agree well with the experimental results of SNWTs fabricated on bulk silicon substrate

7.3 Study on Variability and Reliability of SNWTs

In addition to the parasitic capacitances and self-heating effect discussed in Sect. 7.2, the variability and reliability are two important practical issues for the VLSI application of GAA SNWTs. In this section, the variability, reliability, and noise characteristics of SNWTs will be discussed in-depth.

7.3.1 Experimental and Theoretical Study on Variability of SNWTs

7.3.1.1 Key Variation Sources in SNWTs

The main variation sources in SNWTs are schematically shown in Fig. 7.15 [13]. First of all, SNWTs have several same conventional variation sources as in planar devices, such as metal-gate work function variation (WFV) due to limited size and different orientations of metal grains, the gate length (L_g) variation due to the fluctuations of lithography and etching process of gate pattern, and the variation of carrier transport property due to strain variation and scattering fluctuations such as surface roughness, etc. Fortunately, due to the undoped channel, SNWTs do not suffer much from the random dopant fluctuation (RDF) in the channel, which is the dominant and severe variation in conventional planar devices. However, SNWTs have other new or specific variation sources, as discussed below.

As shown in the figure, the first one is the nanowire cross-sectional shape variation: the nanowire radius (R) can fluctuate, and the nanowire could not be perfect round so that it may more like ellipse or rectangle. The second one is nanowire

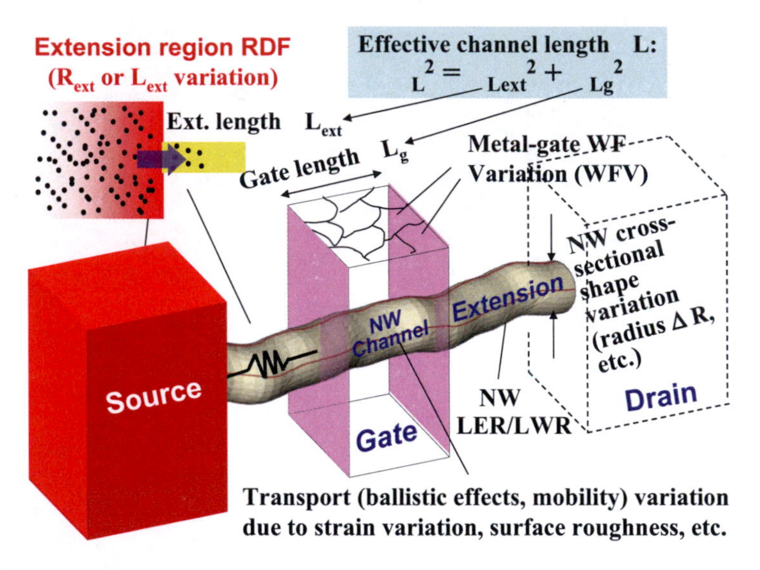

Fig. 7.15 The schematic view of variation sources of SNWT

lateral shape variation, i.e., the nanowire line-edge/line-width roughness (LER/LWR), which allows the nanowire edges varying in arbitrary transverse direction. This nanowire LER mainly originates from the nanowire formation processes (e.g., initial fin patterning process, nanowire size reduction process, and special processes like self-limiting oxidation or H_2 annealing also influence the LER property). The other one is the RDF in nanowire source/drain extension (SDE) regions, although SNWTs do not have severe channel RDF problem. This SDE RDF issue originates from the variations of source/drain implantation and annealing: although people do not dope the nanowires, impurities from the source/drain region can diffuse into the nanowire channel to dope the SDE region (as schematically shown in the inserted figure of Fig. 7.15), resulting in the variations of dopants' number and positions in SDE. This effect will result in the variation of parasitic extension resistance (R_{ext}) and the extension length (L_{ext}).

Among all of the above variation sources in SNWTs, the nanowire LER/LWR is unique [14], which will be studied in more details below.

7.3.1.2 Investigations on Nanowire LER/LWR

Conventionally, there are two different methods to investigate LER/LWR: one is to investigate LER on both sides of the line, and the other is to investigate LWR instead. However, neither of the two methods is sufficient for the description of the lateral shape variation. In the former method, the cross-correlation information of two edges is ignored; while in the latter way, the synchronous information of two edges is missing. Therefore, we have proposed a theoretical model to take into account the correlation between LER and LWR, providing a more accurate description for the lateral shape variation [15]. The model is based on the characterization methodology of auto-correlation function (ACF). In this model, LWR ACF has two components: one is LER ACF and the other is LER cross-correlation function (CCF). Other than conventional LER/LWR parameters of amplitude (Δ) and auto-correlation length (Λ), additional characteristic parameters, the correlation coefficient (ρ) and the translation length (ξ), are adopted to represent the missing cross-correlation information in conventional approaches of LER/LWR description. Generally speaking, the correlation coefficient (ρ) of the two LERs can be considered as the amplitude of the cross-correlation of the two edges, reflecting how strong the correlation is; the translation length (ξ) can be considered as the "period" of the cross-correlation of the two edges, reflecting how wide the highly correlated region is. Therefore, our model can provide the most complete information of LER/LWR.

In order to study the process dependence of LER/LWR in experiments, fins and nanowires are patterned by three different techniques, including hard mask trimming (HT), SiN spacer define (SD), and e-beam lithography (EBL). And nanowire channels are formed by self-limiting oxidation of the initial fins to get well controlled cross-sections, as discussed in Sect. 7.2.1. The key results are shown in Figs. 7.16 and 7.17 [16].

The results indicate that both fins and nanowires have strongly correlated LER/LWR, and the cross-correlation of two edges depends on the fabrication process. In addition, self-limiting oxidation process is found to be helpful to increase correlation length and reduce Gaussian component in LER/LWR ACF.

Furthermore, the impacts of correlated LER on device electrical characteristics are studied by advanced device simulations [15]. One of the typical results is shown in Fig. 7.18. It indicates that V_{th} distribution has strong dependence on the cross-correlation between LER edges, which was missing in previous studies. Non-Gaussian distribution is observed, which shows a much larger variation than that in

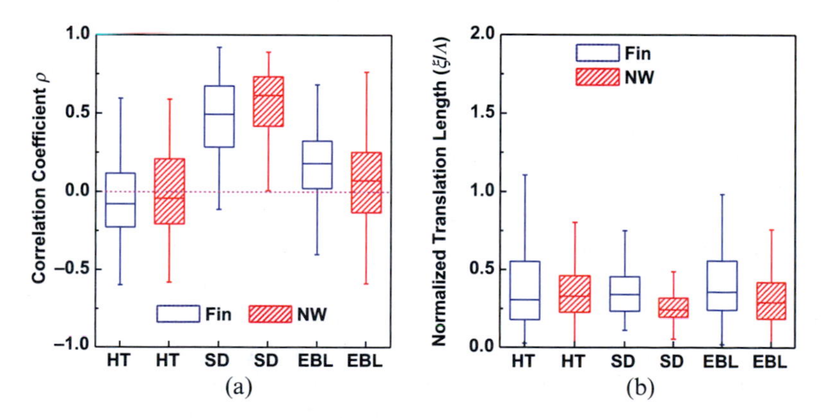

Fig. 7.16 (**a**) Box chart of correlation coefficient ρ. (**b**) Box chart of normalized translation length (ξ/Λ_{eff}) of Fin/NW LER by different fabrication processes

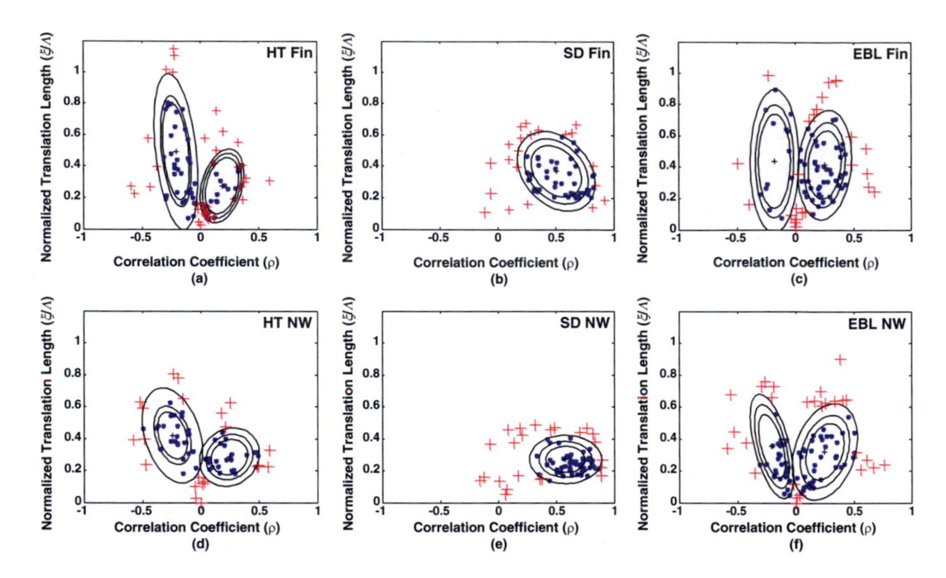

Fig. 7.17 Normalized translation length (ξ/Λ) vs. correlation coefficient ρ extracted from experimental results of Fin/nanowire under different patterning techniques: (**a**, **d**) hard mask trimming (HT), (**b**, **e**) spacer define (SD) and (**c**, **f**) e-beam lithography (EBL). Fiducial confidence ellipses are drawn in the picture with confidence level of 0.3, 0.5, and 0.7

conventional prediction. As a result, the LER effect could be under-estimated if the cross-correlation of LERs is not taken into account.

7.3.2 Experimental Extraction of Variations in SNWTs

The experimental variability characterization methodology follows the measurement-modeling reverse approach [13]. The V_{th} and I_{on} variations are measured in SNWTs with different L_g. In order to reproduce the experimental results, HSPICE Monte Carlo simulation with 1000 runs per experiment is performed. Figure 7.19 illustrates that the agreement of measured and simulated data for both V_{th} and on-state current variations can be achieved simultaneously.

The characterization results show that R and L variations in SNWTs are about 1 nm and 9 nm, respectively. The nanowire LWR amplitude is about 1 nm and correlation length is about 20 nm. The TiN metal-gate WFV is about several decades of mV and tends to increase with scaled L_g. As expected, the L variation is relatively large due to the additional contribution of L_{ext} variation. The absolute value of the variations obtained above may not be that important, because the devices measured in this work are not fabricated with variability-aware optimized processes (so that the results could be treated as the upper limit of SNWT variations) and thus can be

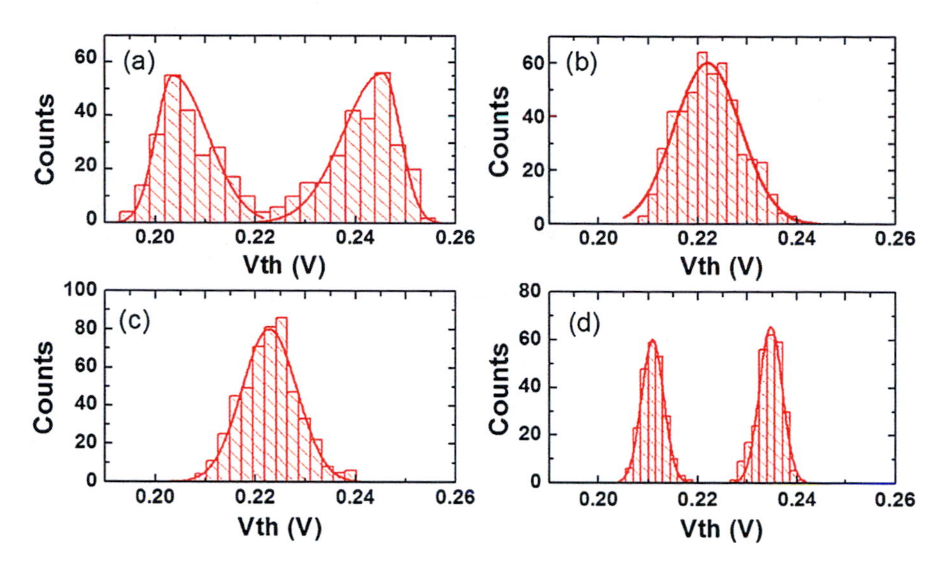

Fig. 7.18 Distributions of threshold voltage with $\Lambda/L_g = 0.5$ under four types of cross-correlation: (**a**) $\rho = -0.5$ and $\xi/\Lambda = 0$, (**b**) $\rho = 0.5$ and $\xi/\Lambda = 0$, (**c**) $\rho = -0.5$ and $\xi/\Lambda = 0.5$, (**d**) $\rho = 0.5$, $\xi/\Lambda = 0.5$

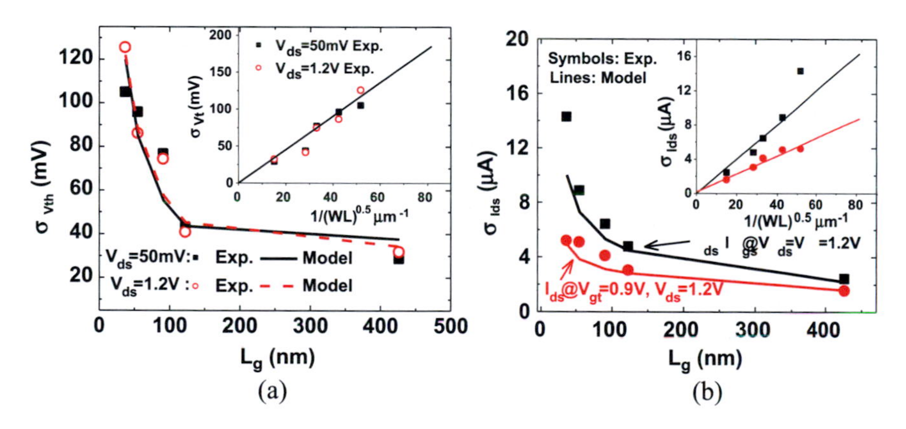

Fig. 7.19 (**a**) Comparison of measured V_{th} variation as a function of L_g and the prediction by HSPICE Monte Carlo simulation of 1000 runs. (**b**) The measured and calculated L_g dependence of I_{ds} variation at $V_{gs} = V_{ds} = 1.2$ V and $V_{gt} = 0.9$ V with $V_{ds} = 1.2$ V

further improved. The critical issue is how much of these sources contribute to SNWT variability compared with each other. Therefore, Fig. 7.20 presents the contributions of different variation sources in SNWTs as a function of L_g. The results indicate that nanowire radius variation and metal-gate work function variation (WFV) dominate both the threshold voltage and on-state current variations due to the ultra-scaled dimensions and strong quantum confinement effects in GAA nanowire structure. The nanowire LER also contributes to the threshold voltage

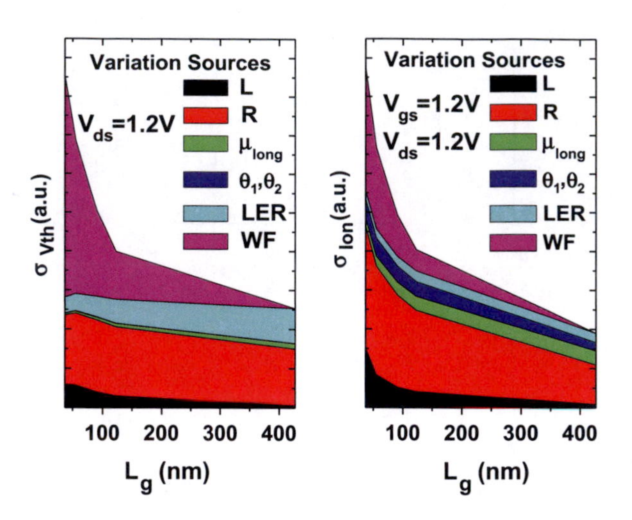

Fig. 7.20 Measured contributions of the main variation sources to V_{th} and ON current variations in SNWTs as a function of L_g

variation, which cannot be ignored. The results provide helpful guidelines for variability optimization of SNWTs [17].

7.3.3 Experimental Study on Reliability of SNWTs

The most important degradation mechanism of transistor reliability in advanced CMOS technology is negative bias instability (NBTI). Our experimental results indicate that the reliability issues of SNWTs exhibit some new features unique to the quasi-1D channel and 3-D surrounding gate [18, 19]. The intrinsic (or average) NBTI characteristics of SNWTs show fast initial degradation, quick degradation saturation, and bias dependent of recovery with more oxide hole detrapping at $V_G > 0$, as shown in Fig. 7.21. These are due to the structural nature of nanowire devices. The GAA structure results in remarkable enhancement of the electrical field near the channel surface, due to the large curvature of the concentric cylinder capacitance, which accelerates the oxide hole trapping effect. It can be further enhanced by the strain in gate oxide induced during the self-limiting oxidation for nanowire thinning and shaping. In addition, the cross-sectional geometry effect of the cylinder wire can also cause faster trap generation at the initial stages. In the SNWT with small gate-area, the thin gate oxide only includes a small number of oxide defects, which can result in quick saturation. In addition, in some SNWTs the impact of electron trapping detrapping on the NBTI behavior is observed [20], which is usually neglected in the traditional devices.

The statistical NBTI characteristics in ultra-scaled SNWTs exhibit fluctuation during degradation with enhanced single and few trap behavior [19, 21], as shown in Fig. 7.22. This NBTI fluctuation mainly originates from the ultra-small gate-area of short-channel SNWTs and the statistical nature of randomly trapped charges. The

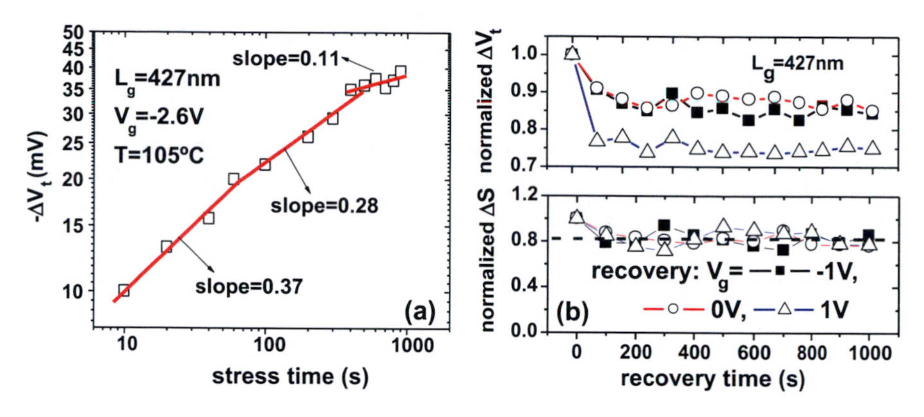

Fig. 7.21 Typical results of (**a**) degradation and (**b**) recovery of NBTI in p-SNWTs

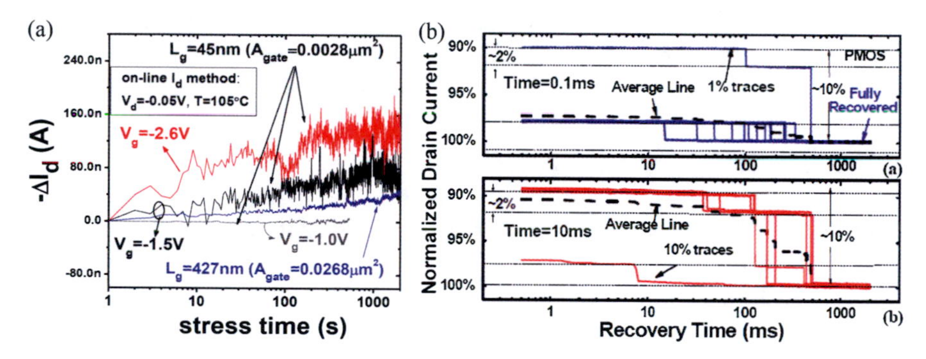

Fig. 7.22 (**a**) Typical results of NBTI degradation in short-channel SNWTs. (**b**) The typical "statistical trap response" (STR) measurement results in SNWTs under different stress time, showing single/few trap detrapping behavior

gate-area of short SNWTs is ultra-small, thus only few trapped charges in the oxide. The random dynamic variations in the number and spatial distribution of the trapped charges can affect the drain current, depending on the local current density in its percolation path through the channel. While, in large-area planar devices, this random nature is hidden by average effects. Therefore, how to characterize the NBTI in ultra-scaled SNWTs is challenging. We have proposed two methods, one is an on-line Ig method [17] which can effectively suppress the NBTI fluctuation of short-channel SNWTs, the other is "statistical trap response" (STR) method [21, 22], which can be used to directly study the impacts of single/few trap in SNWTs during NBTI stress. More details can be found in [23].

Therefore, in addition to further in-depth analysis, special device-circuit co-design methodology or trap-aware design is highly required to maximize the inherent advantages of SNWTs.

7.3.4 Experimental and Theoretical Study on NOISE of SNWTs

In this part, we will discuss two types of low-frequency noise in SNWTs: one is the flicker noise (or 1/*f* noise) in frequency domain, which is important for the analog/RF applications; the other one is the random telegraph noise (RTN) in time domain, which is important for digital applications.

A. Flicker Noise in SNWTs

As shown in Fig. 7.23, the drain current spectral density of ultra-scaled SNWTs exhibits significant dispersion of up to five orders of magnitude [24].

The measured results show that low-frequency noise in SNWTs can be well described by the correlated-mobility fluctuation model at low drain current. As shown in Fig. 7.24, at high drain current, however, the input-referred noise spectral density increases rapidly with the drain current, which indicates the significant impact of the ultra-narrow source/drain extension regions of SNWTs [24]. As a result, design optimizations to reduce the impact of parasitic resistance in SNWTs are necessary for analog/RF applications.

B. Random Telegraph Noise (RTN) in SNWTs

Random telegraph noise is the special case of the low-frequency noise in the time domain, which manifests itself as a random switching of the device current between two or several states due to capture/emission of carriers by localized oxide traps (named border traps or switching oxide trap and other names in the history). Since the RTN amplitude increases rapidly with device scaling, it will strongly impact the

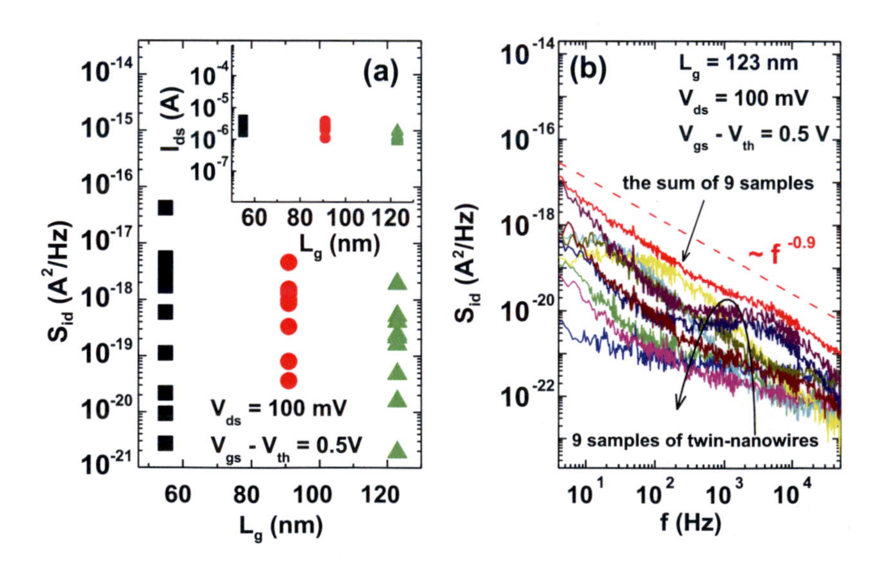

Fig. 7.23 (**a**) Measured dispersion of drain current spectral density of SNWTs at f = 10 Hz, V_{ds} = 100 mV, $V_{gs} - V_{th}$ = 0.5 V. The insert shows measured drain current dispersion of SNWTs at V_{ds} = 100 mV, $V_{gs} - V_{th}$ = 0.5 V. (**b**) Drain current spectral density of 9 individual twin Si nanowire MOSFETs with L_g = 123 nm biased at V_{ds} = 100 mV, V_{gs}–V_{th} = 0.5 V and their sum

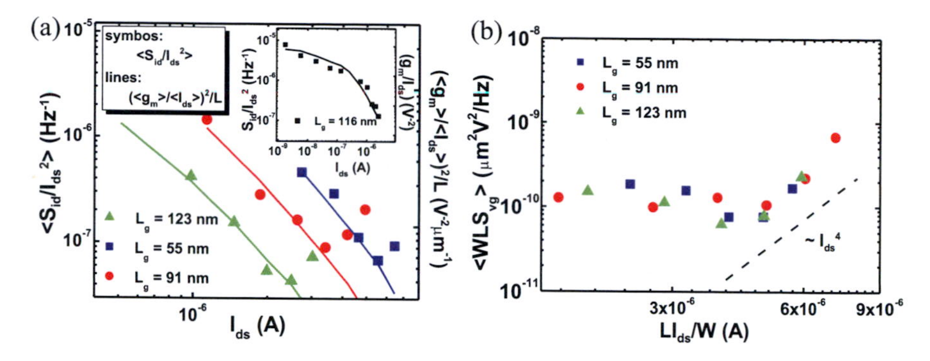

Fig. 7.24 (**a**) Average normalized drain current spectral density $<S_{id}/I_{ds}^2>$ and $(<g_m>/<I_{ds}>)^2/L$ at V_{ds} = 100 mV and f = 10 Hz versus the drain current, for devices with the gate length of 55 nm, 91 nm, and 123 nm. The measured data from subthreshold region to strong inversion region is also shown in the insert. (**b**) Variations of the average normalized input-referred noise spectral density $<WLS_{vg}>$ as a function of the average normalized drain current $<LI_{ds}/W>$ at V_{ds} = 100 mV, f = 10 Hz

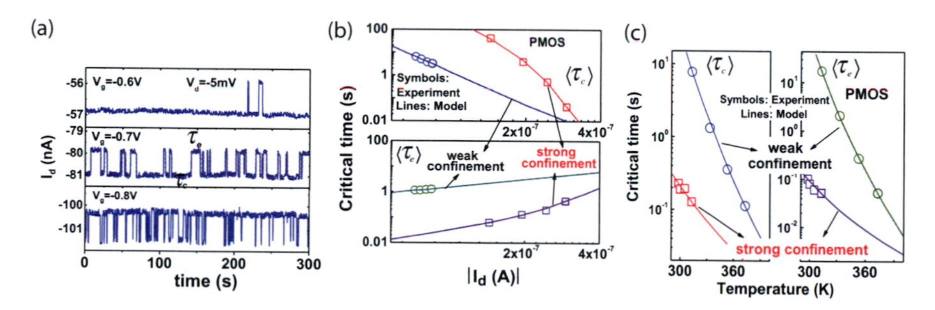

Fig. 7.25 (**a**) Typical measured RTN in a p-type SNWT with d_{NW} = 10 nm, L_g = 58 nm and V_d = 5 mV. (**b**) Typical experimental results (symbols) of the measured capture time constant τ_c and emission time constant τ_e with varying drain current (i.e., gate bias) of RTN in SNWTs with large diameter (~80 nm for weak confinement) and small diameter (~10 nm for strong confinement). (**c**) Typical experimental results (symbols) of the measured τ_c and τ_e with varying temperature of RTN in SNWTs with large diameter (~80 nm for weak confinement) and small diameter (~10 nm for strong confinement)

device reliability and circuit stability. The general review on RTN in advanced logic technology, from device, circuit, and EDA perspectives, can be found in reference [25]. Here we only focus on the different RTN behavior in SNWTs [21, 26].

We have found that the RTN statistical properties in GAA SNWTs are strongly impacted by the quantum confinement effects. As shown in Fig. 7.25, the experimental results indicate that the strong quantum confinement in SNWTs enhances the bias dependence but alleviates the temperature dependence of RTN capture/emission time constants, which cannot be explained by classical RTN model based on Shockley-Reed-Hall (SRH) theory. Thus the lattice relaxation and multi-phonon processes should be taken into account. In order to fundamentally understand the impacts of quantum confinement on RTN time constants, we have developed a full quantum RTN model for SNWTs [21], which agrees well the experimental results. It is helpful for the resilient circuit design of SNWTs against RTN. More details can be found in [27].

7.4 Technology Diversity of SNWTs

It has been controversial for a long time if SNWTs can be feasible or not in most areas where conventional planar MOSFETs have worked successfully, such as I/O, RF/Analog circuit, and so on. Actually, not only as the core devices in VLSI, can SNWTs also play more important roles in other technology applications. In this section, the technology diversity of SNWTs will be discussed with several experimental and theoretical studies.

7.4.1 Electrostatic Discharging by SNWTs

One of the most concern about the feasibility of SNWTs in VLSI is if it can protect the core circuit by effectively electrostatic discharging. Some research work seems to sentence SNWTs to death as an electrostatic discharging device (ESD) due to low thermal conductivity of nanowire and low current in single nanowire [28]. Experiments show that the heat accumulation may easily break down the gate oxide and even distort the nanowire itself physically [28]. It has been suggested that in SNWT technology, other type MOSFETs have to be introduced to build the peripheral circuit, which will push up the process complexity and cost undoubtedly. Therefore, to find an optimized way to enhance the ESD performance of SNWTs becomes very important for the cost-effective VLSI application. In this section, we will discuss about the ESD design by electro-thermal analysis and optimization for SNWTs in details.

A careful electro-thermal analysis has been carried just as in Sect. 7.2.4 to find out the most effective heat dissipation path in SNWTs. As shown in Fig. 7.26a, c, it has been found that the nanowire-gate-substrate is the main path to cool down the device, so that increasing gate length and reducing the isolation thickness between nanowire and substrate can help to improve the electro-thermal characteristics of SNWTs. Accordingly, we proposed a modified gate-grounded SNWTs as ESD device with thinning the isolation layer between nanowire and substrate [29], as shown in Fig. 7.26b. The modified SNWTs show obviously enhanced thermal conductivity even better than other emerging devices such as bulk FinFET and UTBB-SOI MOSFETs, as shown in Fig. 7.26d.

Besides thermal conductivity problem, another limitation on ESD performance of SNWTs is the current conducting capability per single nanowire due to the scaled diameter. To undergo very high failure current possibly up to ampere magnitude, tens of thousands of nanowires should be paralleled together. Due to the size variation and possible weak points on some nanowires, an early breakdown may happen during ESD event to cause the protection function fail. That is why the SNWTs are not convincing in I/O circuit. To evaluate the practical ESD protection capability of the modified SNWTs, the in-house fabrication is carried to obtain a multi-finger SNWT with up to 1400 nanowires by the self-limiting oxidation process described in Sect. 7.2.1, as shown in Fig. 7.27. Due to the good size and roughness controlling,

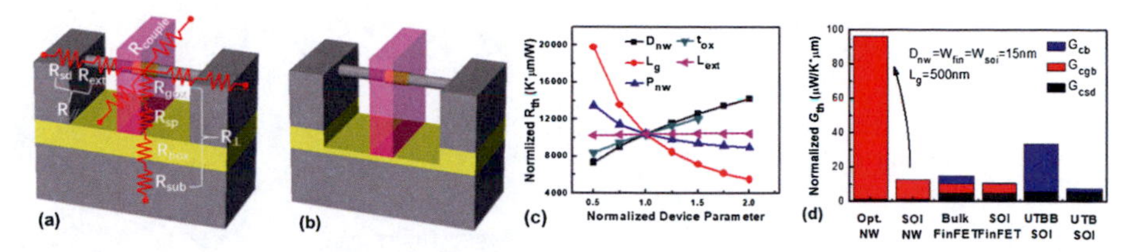

Fig. 7.26 The schematic views of (**a**) thermal resistance network of a normal SNWT and (**b**) a modified SNWT as ESD device, (**c**) the normalized thermal resistance R_{th} as functions of different device parameters, and (**d**) the thermal conductivity of modified SNWTs compared with other emerging devices

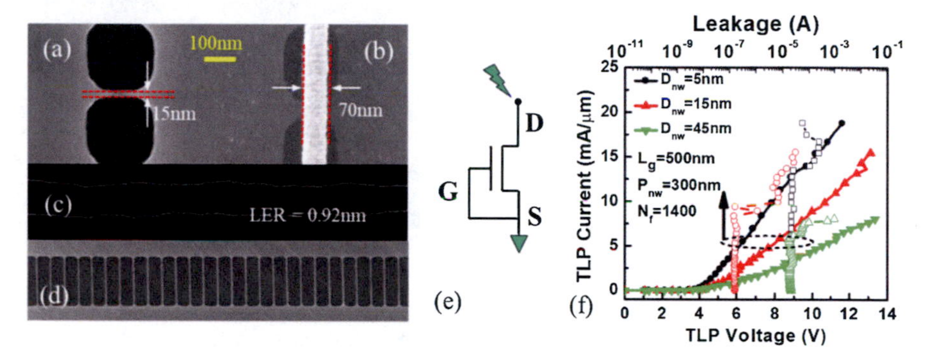

Fig. 7.27 The scan electron microscope images of (**a**) the nanowire formation, (**b**) gate formation, (**c**) line-edge roughness, (**d**) the nanowire array, (**e**) the schematic of DUT for TLP measurement, and (**f**) the measured TLP current characteristics with different nanowire diameters

the smooth and uniform nanowire array is obtained with LER of 0.92 nm. The human-body-model (HBM) transmission line pulse measurement is carried to show the highest failure current of 1.6A at nanowire diameter of 45 nm and gate length of 500 nm. Even as nanowire diameter is scaled down to 5 nm, the failure current can still achieve about 413 mA with the record failure current density of 18.8 mA/μm. The first triggering and the hold voltages are referred to the same point in the SNWTs due to floating body effect. Both of them are measured to be about 4 V, reasonable for avoiding the error triggering. The results show that SNWTs are still feasible in ESD application with optimized thermal dissipation structure.

7.4.2 RF/Analog Circuit with SNWTs

The SNWTs have shown good performance potential in core circuits [30] such as SRAM and ring-oscillator but not clear in mixed-signal and analog integrated circuit application. To verify the feasibility of SNWTs in analog integrated circuit, we have successfully fabricated the high-performance SNWT-based current mirrors (NWCMs) with the two-transistor and cascade circuit configurations on bulk silicon

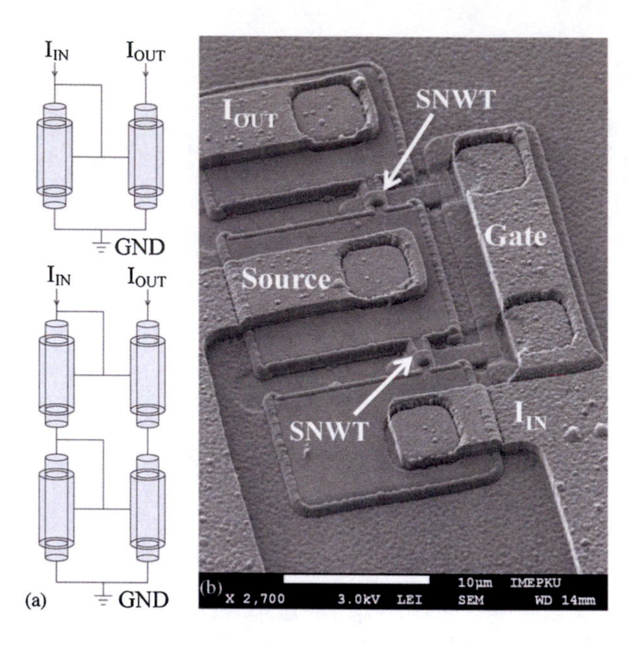

Fig. 7.28 (**a**) Circuit schematics of the (upper) 2T-NWCM and (lower) 4T-CNWCM. (**b**) SEM view of the fabricated 2T-NWCM with a current ratio of 1:1

substrate as shown in Fig. 7.28 [31]. The current ratios CR = I_{out}:I_{in} of NWCMs are set by the nanowire number ratio of output and input transistors.

Figure 7.29 shows the measured current transfer characteristics of three 2T-NWCMs and output characteristics compared the two-transistor planar current mirrors (2T-PCMs). It can be observed that 2T-NWCMs in inversion operation can obtain the expected CR, with input current I_{in} ranging from less than 500 nA to above 10 μA. As the index for the capability of maintaining the prescribed output current under varying V_{out}, the output voltage coefficient OVC(%) = $100(\Delta I_{out}/I_{out})/\Delta V_{out}$ was extracted from the output characteristics in Fig. 7.29. Compared with the average OVC of 5.7% of the 2T-PCM at inversion, which is unacceptable for many analog and mixed-signal applications, the OVC of the 2T-NWCM is significantly low to be within 0.2%. The improvement of OVC can be explained by better suppression of channel length modulation resulting from the improved transport performance in the 1-D channel with GAA structure. With the cascade structure, the OVC of 4T-CNWCM can be further reduced to less than 0.05%.

NWCMs and PCMs in subthreshold operation for ultra-low power applications are also experimentally investigated. The OVC of the subthreshold-biased 2T-PCM can be degraded up to 9.5%; however, no obvious increase in OVC is found in the subthreshold-biased 2T-NWCM (=0.35%), indicating that the NWCM is very attractive for low-power subthreshold circuits compared with traditional planar-MOSFET-based circuits. In addition, the OVC suppression (defined as the OVC suppression of the SNWT-based analog/mixed-signal circuits compared with the planar MOSFETs based circuits) is higher for shorter channel SNWTs circuits due to enhanced SCE suppression of quasi-1D channel and the surrounding gate structure. Particularly nearly, 40% OVC reduction can be obtained for SNWT circuits

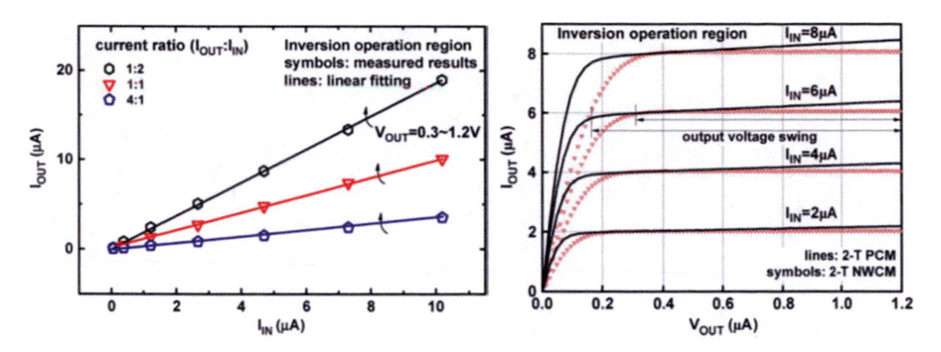

Fig. 7.29 (**a**) Measured current transfer characteristics of three 2T-NWCMs, with CR of 1:2, 1:1, and 4:1, respectively, and (**b**) output characteristics of the 2T-NWCM and the 2T-PCM, with CR of 1:1

with shorter gate length operating at the subthreshold region, which further gives the scaling advantages with subthreshold operation in SNWT-based circuits.

7.4.3 Organic/Silicon Hybrid Integration of SNWTs

Another important application for SNWTs is the sensor of photon, gas, and molecule and so on, due to its large surface-to-volume ratio and very sensitive electrical characteristics to the environmental variables. The conventional way of making a nanowire sensor is to fabricate the nanowires with chemical vapor deposition or catalytic growth in liquid or transferring from other sources such as carbon nanotubes, self-assembling silicon nanowires, and so on. Some researchers also fabricate the nanowire sensor with the top-down approach. No matter bottom-up and top-down scheme, a challenge for nanowire sensor technology is how to assemble the necessary functional materials, usually organic molecules, onto the very limited surface of nanowire with high selectivity. Otherwise, the random distributed organic molecules within the sensing window will cause unpredictable noise to the sensing currents, which are normally at a very low level such as pA or nA.

To solve the selective assembling problem, we have proposed a smart technique to fabricate the organic molecule selectively assembling SNWT sensor [32]. In this scheme, the dielectric materials with different hydrophobicity are used in the sensing window and on the nanowire surface, respectively. By covering the nanowires with silicon dioxide and isolating them from surroundings with silicon nitride, the organic solution will be automatically adsorbed and concentrated onto the surface of nanowire due to the difference in surface tensions on silicon dioxide and silicon nitride. As shown in Fig. 7.30, a porphyrin-silicon hybrid integrated nanowire transistor (PSNFET) is fabricated with such scheme. Porphyrin has been successfully assembled only onto the nanowire surface without any scattered pieces around. However, with the conventional scheme, porphyrin is just assembled in the window at random.

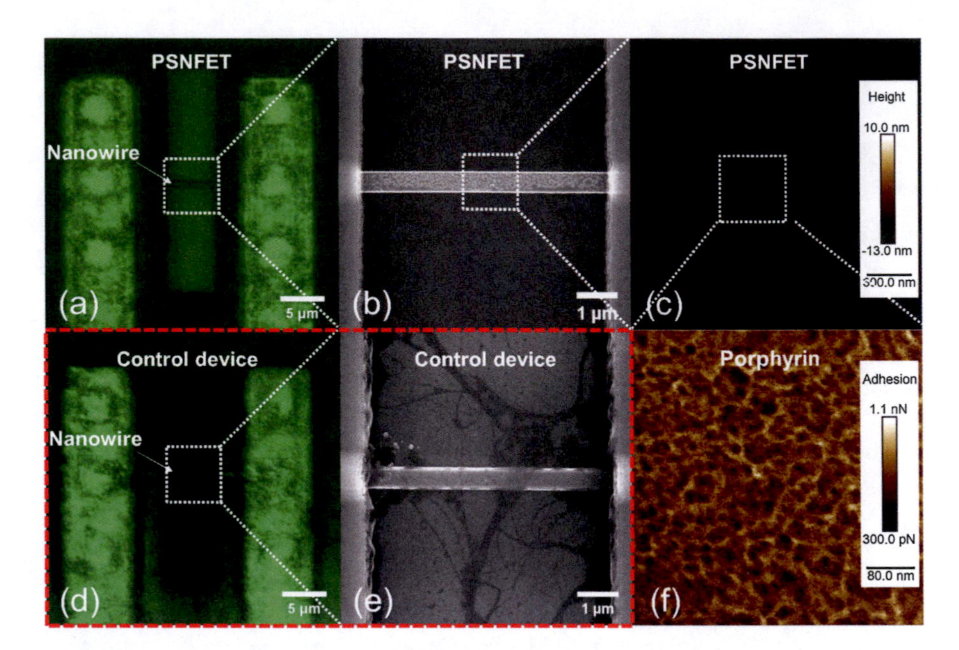

Fig. 7.30 The laser confocal fluorescence microscopy images (**a**, **d**) and scan electron microscope images (**b**, **e**) of the PSNFET and control device, respectively. (**c**, **f**) show the AFM image of PSNFET and the zoom-in view of nanowire, respectively

As a result, the porphyrin molecule is more tightly bonded onto the surface of nanowire through hydrogen bonds and π-π bonds, so that the sensitivity of PSNFET to porphyrin's mole concentration is obviously improved, as shown in Fig. 7.31b. On the other hand, the porphyrin assembled PSNFET can be sensitive to the light incident and thus function as a good photonic sensor due to the photon-induced electron transferring between nanowire conduction band and the LUMO levels in porphyrin molecule. Also in Fig. 7.31c, d shown is the sensitivity comparison of PSNFET and control device to the light incidence. It can be found that even with 10 μM porphyrin solution PSNFET still shows much better sensitivity to the light than the control device with one order higher magnitude of porphyrin assembling.

With the selective assembling scheme, more sensitive and smarter design for the sensor application of SNWTs will be achieved, which is potential for the future internet-of-things technology.

7.5 Summary

Due to the ideal gate controllability and excellent transport performance as well as the extensive technology diversity, SNWTs have risen up as one of the most potential devices at the end of scaling roadmap. The main challenges facing to SNWTs are the process control, variation/reliability-aware design, and the feasibility beyond

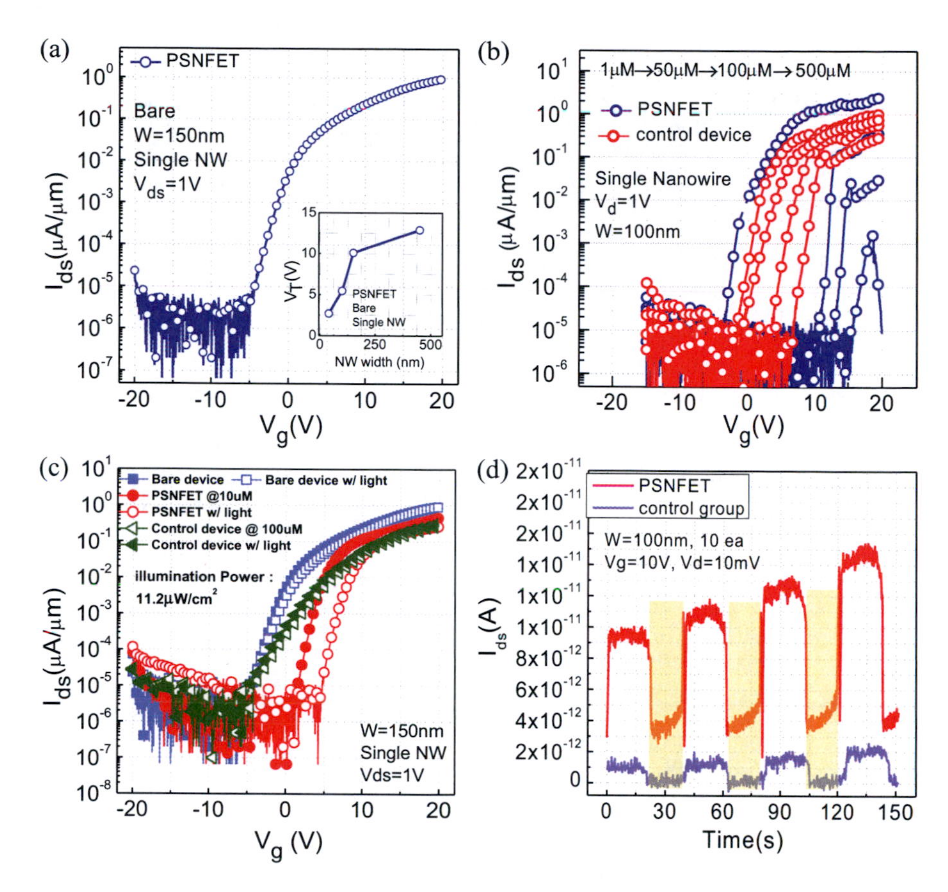

Fig. 7.31 (**a**) Transfer curves of bare PSNFET with single nanowire width of 150 nm. (**b**) Impact of porphyrin molarity on transfer characteristics of PSNFET. (**c**) Transfer curves of bare, porphyrin-coated PSNFET and control device with and w/o light on. (**d**) The sensitivity of 50 μM porphyrin-coated PSNFET and control device to illumination power density of 11.2 μW/cm^2

core circuits. Our recent work has shown the great possibility to solve such problems for SNTWs with device-process and device-circuit co-design. On the other hand, the abundant functions provide SNWTs more values to wider application areas beyond VLSI.

References

1. Huang R, Wang R, Zhuge J, Liu C, Yu T, Zhang L, Huang X, Ai Y, Zou J, Liu Y, Fan J, Liao H, Wang Y (2011) Characterization and analysis of gate-all-around Si nanowire transistors for extreme scaling. In: IEEE Custom Integrated Circuits Conference (CICC), pp 1–8
2. Yu T, Huang R, Wang Y, Zhuge J, Wang R, Liu J, Zhang X, Wang Y (2007) New self-aligned silicon nanowire transistors on bulk substrate fabricated by epi-free compatible CMOS tech-

nology: process integration, experimental characterization of carrier transport and low frequency noise. In: IEEE International Electron Devices Meeting. Technical Digest, pp 895–898
3. Fan J, Huang R, Wang R, Xu Q, Ai Y, Xu X, Li M, Wang Y (2013) Two-dimensional self-limiting wet oxidation of silicon nanowires: experiments and modeling. IEEE Trans Electron Devices 60(9):2747–2753
4. Lundstrom MS, Guo J (2006) Nanoscale transistors: device physics, modeling and simulation. Springer, New York
5. Chen M-J, Huang H-T, Huang K-C, Chen P-N, Chang C-S, Diaz CH (2002) Temperature dependent channel backscattering coefficients in nanoscale MOSFETs. In: Digest. International Electron Devices Meeting. Technical Digest, pp 39–42
6. Wang R, Liu H, Huang R, Zhuge J, Zhang L, Kim D-W, Zhang X, Park D, Wang Y (2008) Experimental investigations on carrier transport in Si nanowire transistors: ballistic efficiency and apparent mobility. IEEE Trans Electron Devices 55(11):2960–2967
7. Wang R, Zhuge J, Liu C, Huang R, Kim D-W, Park D, Wang Y (2008) Experimental study on quasi-ballistic transport in silicon nanowire transistors and the impact of self-heating effects. In: IEEE International Electron Devices Meeting. Technical Digest, pp 1–4
8. Huang R, Wang R (2010) Investigation of gate-all-around silicon nanowire transistors for ultimately scaled CMOS technology from top-down approach. Front PhysChina 5(4):414–421
9. Zhuge J, Wang R, Huang R, Zhang X, Wang Y (2008) Investigation of parasitic effects and design optimization in silicon nanowire MOSFETs for RF applications. IEEE Trans Electron Devices 55(8):2142–2147
10. Zou J, Xu Q, Luo J, Wang R, Huang R, Wang Y (2011) Predictive 3-D Modeling of parasitic gate capacitance in gate-all-around cylindrical silicon nanowire MOSFETs. IEEE Trans Electron Devices 58(10):3379–3387
11. Wang R, Zhuge J, Huang R, Kim D-W, Park D, Wang Y (2009) Investigation on self-heating effect in gate-all-around silicon nanowire MOSFETs from top-down approach. IEEE Electron Device Lett 30(5):559–561
12. Huang X, Zhang T, Wang R, Liu C, Liu Y, Huang R (2012) Self-heating effects in gate-all-around silicon nanowire MOSFETs: modeling and analysis. In: Thirteenth International Symposium on Quality Electronic Design (ISQED), pp 727–731
13. Wang R, Zhuge J, Huang R, Yu T, Zou J, Kim D-W, Park D, Wang Y (2011) Investigation on variability in metal-gate Si nanowire MOSFETs: analysis of variation sources and experimental characterization. IEEE Electron Device Lett 58(8):2317–2325
14. Wang R, Yu T, Huang R, Ai Y, Pu S, Hao Z, Zhuge J, Wang Y (2010) New observations of suppressed randomization in LER/LWR of Si nanowire transistors: experiments and mechanism analysis. In: 2010 International Electron Devices Meeting. Technical Digest, pp 34.6.1–34.6.4
15. Jiang X, Wang R, Yu T, Chen J, Huang R (2013) Investigations on line-edge roughness (LER) and line-width roughness (LWR) in nanoscale CMOS technology: part I – Modeling and simulation method. IEEE Electron Device Lett 60(11):3669–3675
16. Wang R, Jiang X, Yu T, Fan J, Chen J, Huang R (2013) Investigations on line-edge roughness (LER) and line-width roughness (LWR) in nanoscale CMOS technology: part II – experimental results and impacts on device variability. IEEE Electron Device Lett 60(11):3676–3682
17. Zhuge J, Wang R, Huang R, Zou J, Huang X, Kim D-W, Park D, Zhang X, Wang Y (2009) Experimental investigation and design optimization guidelines of characteristic variability in silicon nanowire CMOS technology. In: 2009 IEEE International Electron Devices Meeting (IEDM). Technical Digest, pp 61–64
18. Huang R, Wang R, Liu C, Zhang L, Zhuge J, Yu T, Zou J, Liu Y, Wang Y (2011) HCI and NBTI induced degradation in gate-all-around silicon nanowire transistors. Microelectron Reliab 51(9–11):1515–1520
19. Wang R, Huang R, Kim D-W, He Y, Wang Z, Jia G, Park D, Wang Y (2007) New observations on the hot carrier and NBTI reliability of silicon nanowire transistors. In: IEEE International Electron Devices Meeting - Washington, DC, pp 821–824

20. Zhang L, Wang R, Zhuge J, Huang R, Kim D-W, Park D, Wang Y (2008) Impacts of non-negligible electron trapping/detrapping on the NBTI characteristics in silicon nanowire transistors with TiN metal gates. In: International Electron Devices Meeting. Technical Digest, pp 123–126
21. Liu C, Wang R, Zou J, Huang R, Fan C, Zhang L, Fan J, Ai Y, Wang Y (2011) New understanding of the statistics of random telegraph noise in Si nanowire transistors – the role of quantum confinement and non-stationary effects. In: International Electron Devices Meeting. Technical Digest, pp 521–524
22. Liu C, Zou J, Wang R, Huang R, Xu X, Liu J, Wu H, Wang Y (2011) Towards the systematic study of aging induced dynamic variability in nano-MOSFETs: adding the missing cycle-to-cycle variation effects into device-to-device variation. International Electron Devices Meeting. Technical Digest, pp 571–574
23. Huang R, Wang R, Li M (2014) Characteristics of NBTI in multi-gate devices for highly-scaled CMOS technology. In: Grasser T (ed) Bias temperature instability for devices and circuits. Springer, New York
24. Zhuge J, Wang R, Huang R, Yu T, Zhang L, Kim D-W, Park D, Wang Y (2009) Investigation of low frequency noise in silicon nanowire MOSFETs. IEEE Electron Device Lett 30(1):57–60
25. Wang R, Guo S, Zhang Z, Wang Q, Wu D, Wang J, Huang R (2018) Too noisy at the bottom? —random telegraph noise (RTN) in advanced logic devices and circuits. International Electron Devices Meeting. Technical Digest, pp 388–391
26. Zhuge J, Zhang L, Wang R, Huang R, Kim D-W, Park D, Wang Y (2009) Random telegraph signal noise in gate-all-around silicon nanowire transistors featuring coulomb-blockade characteristic. Appl Phys Lett 94(8):083503
27. Wang R, Liu C, Huang R (2013) Random telegraph noise in multi-gate FinFET/nanowire devices and the impact of quantum confinement. In: Han W, Wang Z (eds) Towards quantum FinFET. Springer, New York
28. Liu W, Liou JJ, Jiang Y, Navab S, Lo GQ, Chung J, Jeong YH (2010) Failure analysis of Si nanowire field-effect transistors subject to electrostatic discharge stresses. IEEE Electron Device Lett 31(9):915–917
29. Li M, Fan J, Xu X, Huang R (2017) Investigation on electrostatic discharge robustness of gate-all-around silicon nanowire transistors combined with thermal analysis. IEEE Electron Device Lett 38(12):1653–1656
30. Liu Y, Huang R, Wang R, Zhuge J, Xu Q, Wang Y (2012) Design optimization for digital circuits built with gate-all-around silicon nanowire transistors. IEEE Electron Device Lett 59(7):1844–1850
31. Huang R, Zou J, Wang R, Fan C, Ai Y, Zhuge J, Wang Y (2011) Experimental demonstration of current mirrors based on silicon nanowire transistors for inversion and subthreshold operations. IEEE Electron Device Lett 58(10):3639–3642
32. Chen G, Yu B, Li X, Xu X, Li Z, Huang R, Li M (2019) Selective-assembling hybrid porphyrin-silicon nanowire field-effect transistor (PSNFET) for photonic sensor. IEEE Electron Device Lett 40(5):812–814

Chapter 8
Heterojunction Bipolar Transistors and Monolithically Integrated Photoreceivers Among Other Applications

Leda Lunardi

8.1 Background

Growing up in a large family in Brazil shaped my personality profoundly. Not only the first generation able to attend college but also to earn graduate degrees. Unfortunately, most of my migrant elderly relatives, parents, and grandparents separated from their original land and families, never could appreciate their grandchildren achievements. The last time I bad farewell to my maternal grandfather who I was very close before leaving to attend graduate school, it was the last time I saw him. He knew I did not.

8.2 Education

I always loved going to school: everything was interesting there, however, because of the rigid catholic discipline my goal was not to get into any trouble. That changed when I transferred to a public school in high school. Yet I never considered a graduate degree because no one in my family had gone to college. At home, careers meant acquiring a profession; most admiration and success was associated with men and their professions. School was reserved for the men because a profession would provide for their future families. There were few exemptions inside my family. Women attending college seemed a waiting game before getting married. Nevertheless, others wanted to study or wish they had the opportunity to. For example, my maternal grandmother. She was illiterate, who never had a chance to improve herself and worked as a blue collar. However, she often looked for opportunities on her life.

L. Lunardi (✉)
Department of Electrical and Computer Engineering, North Carolina State University, Raleigh, NC, USA
e-mail: leda_lunardi@ncsu.edu

A. C. Parker, L. Lunardi (eds.), *Women in Microelectronics*, Women in Engineering and Science, https://doi.org/10.1007/978-3-030-46377-9_8

When her oldest grandchild started primary school, she learned from him reading, writing and the arithmetic that she needed to start a small business of her own. Her first act of freedom was to obtain a driver's license and also to buy an old "stick shift" Jeep; so she could take her husband and the family dog to the seashore while infuriating all other drivers along the way with her slowly driving.

8.2.1 Elementary/High School

In the medical sciences track high school, I had followed my girlfriends from my days of the catholic school, since some intended to attend medical school. Before registering for the required national admissions test in senior year, I decided to make an appointment at one of the best Medical Universities. (Disclosure: the universities in Brazil follow the European system where medicine is a 6-year undergraduate degree; unlike in the American one where law and medicine are 4-year graduate degrees). During the first visit that happened at School of Medicine of University of São Paulo, one of the rooms was the anatomy laboratory. I had an uncontrollable revulsion at the sight of the cadavers. The impact was serious: I had to get out before considering in getting in any school. I had several conversations with my parents and teachers, and started redefining a different major in either languages or psychology. However, my physics teacher, the hardest subject for me at that time, gave me the best advice ever: "You will learn a lot in Physics." Nobody talked to me about engineering. My oldest sister had some friends that were engineering students, but never talked to me about being an engineer because there were very few women engineer students, mainly in civil engineering. Looking back, I try to remember the reasons that the male students would never talk about their studies with the women. In my family, there were no limitations, I could do whatever I wanted, since the best universities were public, if I could get into one. To explain to my grandfather what physics was since he could not understand it, I settle for mathematics and he was happy that one day my children, his great grandchildren would have a mother able to explain their homework.

8.2.2 Physics and Undergraduate Research

As a rising sophomore in Physics, I interviewed and got into an opportunity offered by our instructor in Physics for undergraduate researchers at the Linear Accelerator Laboratory, the prototype of the Mark 3 that Stanford University had donated to Institute of Physics. Following my advisor suggestion, I applied and received an undergraduate research initiation grant from State of São Paulo Agency for Funding Research (FAPESP) to work on my own research project [1]. The project on spectroscopy of radioactive elements was to blast with a few MeV electron beam of radioactive elements' prepared targets such as uranium and thorium. The experimental setup was composed of radiation detectors and sophisticated electronics. Most of the shared equipment had to disassemble after my time slot, and returned in optimum running condition. The calibration of the system on a constant basis was very important, because depending on the cross section of the byproducts, using beta and gamma spectroscopy methods, the lifetimes of the element products and all measurements depended on several precise digits.

8.2.3 Experimental Nuclear Physics

At the time I finished the BS in Physics, most of the experimental research for a M.Sc. thesis was completed. I also continued to accumulate graduate credits to meet the requirements for the PhD in physics [2]. My graduate supervisor was the first PhD woman from the same group I had done all my undergraduate research. She was developing a unified theory on the cross section of delayed neutrons: product of the fission in actinides; however, as new faculty she was under pressure to deliver fast results.

While doing my masters graduate school I realized that experimental nuclear physics was not the career, I wanted to have. There were few options for jobs and teaching as lifetime profession was not on top of my list. I took a year off, started teaching Physics in a private engineering college that was rapidly expanding. In addition to myself, another faculty was full time and had a PhD in electrical engineering. When a chance for a Fulbright Fellowship opportunity came along, I applied with sponsorship by the private college. The PhD colleague helped me suggesting topics for personal plans of work to schools that had strong electrical engineering programs.

This was an incentive to finish the MS thesis and start applying for foreign graduate schools. Even when the Fulbright sponsor fell through as I needed financial support, with the help of former instructors and support of recommendation letters, I was able to secure a full scholarship for a PhD and soon send applications to graduate schools in the USA.

8.2.4 Graduate School in a Foreign Country

When arrived for graduate school at Cornell I felt isolated and as the only woman in several places even if I was already fluent in English. I was surprised with the smaller female student population in both the Physics and Engineering than my original university. One of the instructors of Solid State Physics was surprised when I asked him about the reason there were so few women around. I had to take few physics courses again required for the engineering students. Some like Quantum Mechanics that I had been a teaching assistant. With a full fellowship, I realized that few faculty members were not interested on my background or my plan of work to be part of their group but because I had a fellowship. Therefore, I took control and looked around for the research group that could offer me the best offer. In the first year, I worked with an assistant professor doing ion implant simulations in Silicon structures, and in the summer, I managed to debug a super simulator that had been inactive for a while. At the end, I got an acknowledgment in a footnote in a conference paper. Cornell with the National Science Foundation (NSF) Submicron Facility was one of the top schools for device fabrication. Professor Eastman's research group was the leader in molecular beam epitaxy (MBE) on campus with several visiting scientists in devices. I decided to join his large group, and learn compound semiconductor processing and devices. Students were in either materials or devices. Although in the shared office among the graduate students with thesis topics on devices, I was the only woman. My graduate student colleagues were knowledgeable, spent long hours inside the lab, and very collaborative. I graduated from Cornell University with a PhD in electrical engineering. My dissertation was the first in USA on the topic of AlGaAs/GaAs Heterojunction Bipolar Transistors fabrication and microwave operation, which landed few opportunities to continue my professional career and not to return back home.

8.3 Sponsored Research in Corporate Organizations

When I joined AT&T Bell Laboratories, right after graduating, commercial lightwave systems operated at 500 Mb/s in single mode fiber without wide commercial deployment. One of the main interests in research and development of laboratories was to explore the fundamental limits of electronic devices for optical communication systems [3].

At the Murray Hill location, there were over two hundred epitaxial machines part of the research and development organizations. There was strong research investment on epitaxial growth methods for Ga, In, As, P-based materials to be compatible with 1310 and 1550 nm wavelength applications. It was interesting to have around many epitaxial growers and different options, with physicists, material scientists, engineers, and device experts. There were frequent seminars from researchers of other locations and external visitors.

Photonic devices (such as lasers, photodiodes, modulators) and integration were already been demonstrated. Electronic devices could not fulfill simultaneously challenges such as high frequency performance, minimum noise, high gain, low power, and high level of circuit complexity. None of the Si-based technologies such as bipolar or complementary metal oxide semiconductor (CMOS) or the (then) emerging promising compound semiconductors (GaAs, InP) had the materials requirements and the technology advancements of bipolar-based or field effect transistor (FET)-based to meet the desired 10Gb/s data rate communications requirements.

The next generation of lightwave communications systems (envisioned to be above 10 Gb/s), primarily based on Silicon electronics, would require heterojunction-based transistors. At the time, this was strategic and critical since the main competitors were in R&D laboratories of companies in Japan and Europe.

The job proposition was to develop digital circuits for telecommunication applications as continuation of my PhD work. The group was strong in Gas-based heterojunction FETs with the majority of the laboratory already relocating to Pennsylvania. There was no coordinated activity in compound semiconductor based bipolar technology that could be integrated (different device concepts, isolated demonstrations without integration or demonstrated in circuit simulations!). I would work alone with the research organizations with full support of the management, and a clean room with a FET-based processing line. Maybe I was naïve but I accepted the challenge.

During the interview I met few researchers that later became collaborators, in particularly, R. Malik interested in AlGaAs/GaAs HBT grown by MBE. Our interaction was creative, which produced different transistor structures to minimize the GaAs surface recombination and/or facilitate processing [4].

After giving a seminar in another research location (Holmdel), I met an additional key collaborator, R.G. Swartz, with Silicon CMOS circuit design expertise. I had the critical mass for starting to implement GaAs-based HBTs, and assembled a group of collaborators across three divisions (two in research and one in development, in three different locations) to deliver the building blocks with GaAs-based HBTs digital circuits. During the next year I had put together a high yield technology with MBE epitaxial in house material, the results included small-scale integration with ring oscillators, different designs for frequency dividers, and decision circuits that Bob Swartz for study.

Several R&D laboratories started reporting GaAs-based HBT results, but we were focusing on optical communication systems so the simulations in addition to the standard ring oscillators, frequency dividers, Bob Swartz also included a laser driver and a decision circuit. In digital communications systems, after detection of the optical signal, the conversion from the optical to the electrical domain, the signal is pre-amplified, reconditioned, and reshaped by special amplifiers, followed by a special type of circuit, the decision circuit. After the decision circuit at sampling time from the recovered clock sets the decision for a signal to "1" (one) or a "0" (zero), one of the crucial elements for the signal recovery. Since no one had demonstrated, GaAs-based HBTs decision circuits that was going to be the key contribution [5].

We were the first to demonstrate AlGaAs/GaAs HBT-based decision circuits operating at a frequency higher than 4 Gb/s [6]. Previous reports employing GaAs HFETS were limited to 2 Gb/s including from our own laboratory [7], and NTT in Japan [8]. When reporting the AlGaAs/GaAs HBT-based decision circuit results, my collaborators and I intended to demonstrate that with a modest HBT technology could outperform HFETs. Outside of AT&T in the community, we won the respect especially from NTT-Japan. Other experimental results were never published outside because the edge for having the work done. With detailed simulations, the experimental demonstration with yields over 80% was transferable to a production line similar to the one in the development laboratory. With more established technologies such as GaAs-based HFETs, Si-based BJTs and CMOS, InP-based HBTs with both InGaAs and InAlAs heterojunctions were drawing attention because the compatibility with the long optical wavelength even if integration had yet to be demonstrated. Amid internal tumultuous organizational changes in the project, and without any incentive for my career, I decided to move.

I joined the research laboratory in Crawford Hill, where S. Chandrasekhar was leading the efforts on the InP HBT-based project [9]. For the next 5 years, we would make a strong team, contributing, and leading the world research on long wavelength integrated photo receivers.

Monolithic integration at that time involved small number of optical devices such as lasers, photodetectors, or modulators with electronic circuits resulting in a smaller area, specialized chip that transferred from the optical to the electrical domain and/or vice versa. Photonic integration circuits with only photonic devices passive and active in the optical domain were been demonstrated. The motivation behind the research on optoelectronic integrated circuits (OEICs) was the low noise and the lower component count associated with the integration and in the long range the cost. When compared to hybrid circuits, OEICS that included a photodiode with an HBT-based preamplifier theoretically expected to have improved performance, reduced circuit size and enabling more functions with added circuitry levels. Catalyst for several results and increased awareness for the importance in terms of material development and research. There was some limitations on the doping level for the p-type InP material if the epitaxial was from metal organic vapor deposition (MOCVD); different epitaxial techniques then were not well established for substrates larger than 2″ by MOMBE or MOCVD, and processing development.

My contribution to the project was on the preamplifier design and optimization. For transmission systems, the preamplifier design as a transimpedance type is to transform the small current of the photodiode into an output voltage of a few hundred millivolts peak-to-peak. By using a preamplifier, the main advantages compared to other types of amplifiers (such as a high-impedance one) are the higher dynamic range and the elimination of a critical equalizing circuitry. It was an ideal demonstration of the potential of any technology such as the InP-based HBT.

Furthermore, the Crawford Hill laboratory was a small community with open doors and a collaborative environment. Coming from a large and closed research headquarters that was Murray Hill, it was a fresh change. I appreciated the friendly, professional, and welcoming attitude that everyone had. However, during lunchtime if you decided to play beach volleyball or join the runners, then you would learn their true competitive nature. Nevertheless, science and sense of humor were always at its best.

Chandrasekhar and I started talking about ideas and collaboration into his project. While he invented the integration, processed the chips, I contributed with high-speed characterization of the transistors, developed a device model, designed the amplifiers and the circuits. My first contribution made results accepted in a Post deadline submission [10].

After the initial results, I decided to change to a different preamplifier design with MOMBE material. The measurements were in agreement with my simulations within 10% and had an improvement of 6 dB than our previous results. We obtained a record performance at 10 Gb/s single channel InP-based p-i-n PD HBT-based photoreceivers, with reported sensitivity −20.4 dBm, translated as error free with a non-return-to-zero-ON-OFF-keying (NRZ-OOK) data [11]. The receivers operated up to 20 Gb/s. Furthermore all receivers were packaged and fiber pigtailed confirming that there was no penalty between on wafer and packaging. I was able to test in different setups (the so-called testbeds) around the company, which included terrestrial long distance, fiber loop for cable TV, and submarine among others [12]. These results confirmed that HBT-based OEICs could outperform any other technology.

Compact models are device equations written for a commercial circuit simulation platform and easily incorporated into devices libraries of open access modeling simulators to generate the output characteristics. In collaboration with a visiting scholar, we developed a compact modeling development for InP-based HBT numerical solution for the best fit between the model and the measured DC and microwave data. The simulation of integrated photoreceivers used this model [13].

Later in academia, some of my students would continue to employ compact modeling techniques for extracting data for different types of devices [14].

There were more applications for the OEICs with more channels and other devices: we worked well as a team with Chandrasekhar. With the number of channels expanded to 8- channel p-i-n HBT amplifier array operating and performance at 2.5 GB/s for dense wavelength division multiplexing applications, the aggregated capacity was 20 Gb/s. The 8-channel OEICs became also part of a DARPA-funded project [15]. Another first demonstration was the integration of the 8-channel monolithic integration of photodetector with an amplifier with a waveguide router. The array waveguide router (AWGR) was a monolithic eight-wavelength demultiplexed receiver for dense WDM Applications [16].

All OEIC InP-based HBT research demonstrations were with NRZ formats before the use of high-spectral-efficiency quadrature amplitude modulation necessary to increase both per-channel interface rates and aggregate WDM capacities in communication systems.

Another important and first demonstration for the OEIC technology was with one collaborator from the GaAs-based HBTs team. Bob Swartz and Y. Ota were investigating optical data links and suggested that maybe we should try to make an OEIC for data links [17]. The implementation would be in the form of a DC-coupled differential burst mode photoreceivers unlike conventional AC-coupled optical receivers, intended for continuous data reception and unusable in optical interconnects, and data communication in computer applications. The goal of a burst mode receiver is to improve the efficiency of a data link, achieved by adjusting the transmission data shared among many sources and controlling the signal amplitude, in addition to canceling any dc offset.

The demonstration of an OEIC-based on InP-based HBT burst mode receiver operating s up to 5 Gb/s was the first of its kind with both continuous and burst mode data streams with a sensitivity of −18.6 dBm for the continuous mode operation at a bit error rate of 10^{-9} [18]. Another version, with an improved performance for an integrated photoreceiver incorporates active sources, as displayed in Fig. 8.1.

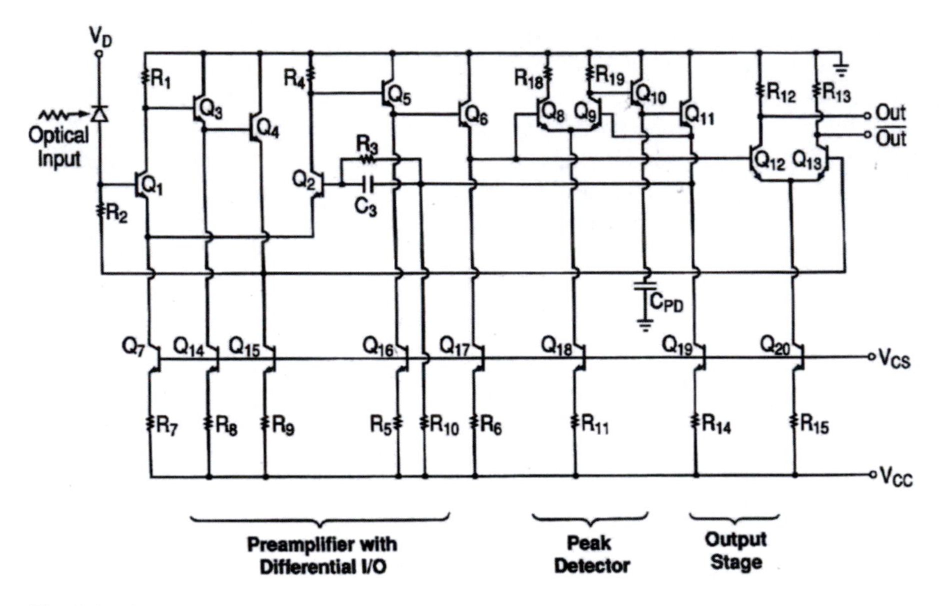

Fig. 8.1 Circuit schematic of the burst mode amplifier with active sources

The optical-to-electrical small-signal measured response of the burst mode receiver shown in Fig. 8.2 with applied bias indicated in the inset.

Measured eye diagrams at 5Gb/s NRZ data streams for both outputs and bias conditions are displayed in Fig. 8.3 below.

The error rate performance was also performed with continuous NRZ 2^{23-1} pattem-length pattern rate bit sequence (PRBS) at 5 Gb/s and 10 Gb/s (Fig. 8.4). Our collaboration demonstrated a wide range of operating speeds from 100 Mb/s to 20 Gb/s for single channel monolithic InP-HBT's photo receivers in optical communications systems in addition to multichannel array receivers with aggregated speeds of 20 Gb/s [15].

With modest technology and superior material, we had reached the maximum speed for single channel operation. More aggressive lithography and device redesign was required in order to meet the need for higher frequencies applications. However, in late September 1995 while attending an optical conference with most of our laboratory, we got the news that AT&T was splitting into three companies. For the next few months, we mourned and started the painful process of separation, with colleagues also retiring or leaving the company. My immediate manager, Marty Pollack, was one of them: the best one I ever had in my whole career to date.

Following the AT&T split into three companies, unlike others without a choice, I decided to stay at AT&T Labs that had a dedicated research lab. The group had to relocate to a temporary place while finding a new location for the new headquarters construction. We started getting set up, learning more about the network, having lectures and seminars from network experts. In the broadband access network group, several approaches on fiber-to-the-home were under investigation. Optical

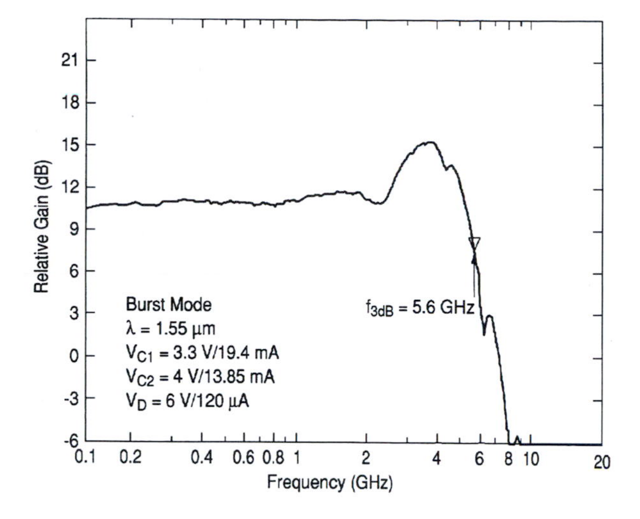

Fig. 8.2 Optical-to-electrical small-signal response of a fiber-pigtailed packaged OEIC burst mode module with a 1.55-μm light source. The bias conditions for the amplifier (V_{C1},V_{C2}) and photodetector (V_D and Iph) are indicated

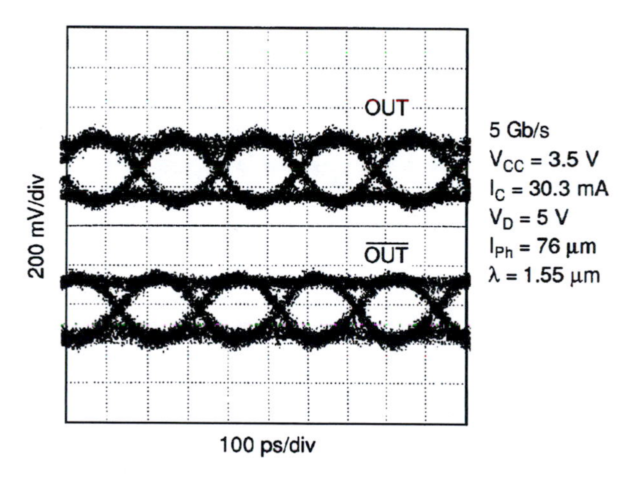

Fig. 8.3 The eye diagram for the two complementary outputs of the burst mode receiver OEIC module measured at 5 Gb/s input data rate with an NRZ 2^{23-1} pattern length PRBS. The inset indicates the bias conditions

communications industry attracted attention from new investment and capital, with new companies forming and changing the landscape. Arriving at work it was the daily surprise of finding out who was pursuing new ventures. For a few years, I could do it but it was not the same after Crawford Hill. The startup fever had hit the telecom industry so I also decided to take a chance and to join JDS Uniphase.

Setting up a small research group inside a fast growing company without traditional research was a completely different experience. The company had a different culture, with conglomerate of different acquired companies: fast pace without paperwork, shallow managerial structure. Experts on each one of the products were available to us. One of the goals of the research group was to build a 40Gb/s optical testbed with JDSU equipment as a showcase. One of the initial challenges was staffing the group. The market was competitive because of the sprawling of startups and expansion of several companies. As the group acquired more members includ-

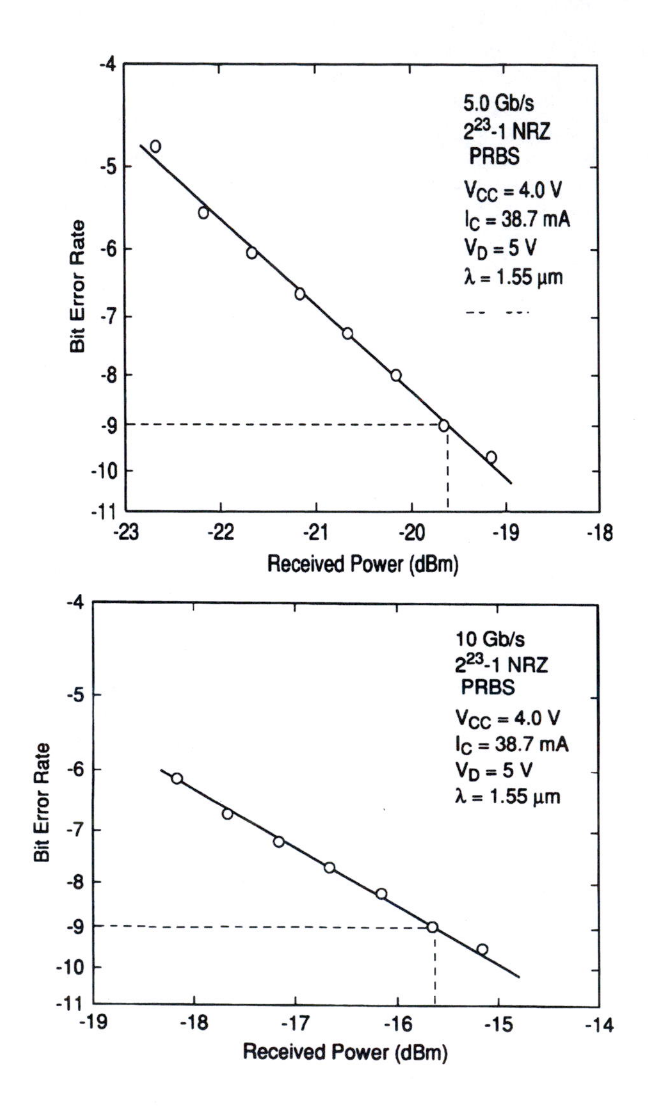

Fig. 8.4 Measured bit error rate as function of the optical incident power at: (top) 5 Gb/s and (bottom) 10 Gb/s with an NRZ 2^{23-1}pattem-length PRBS. The bias conditions are the same as in Fig. 8.3

ing one manager, I receded back to the technical track: title like "Senior Scientist or Technical Leader" were fine to me. The group produced the first 40 Gb/s electronic eye diagram and started contributing to the main optical conferences. One of those devices was an etalon used for dispersion compensation, which gave me an opportunity to collaborate again with Chandra at Lucent Bell Labs [19].

Capacity increase was the result of the technical progress in high-speed electronics (including BiCMOS and compound semiconductors) and the natural progression to meet the demand for communication bandwidth cost effectively. Wavelength Division Multiplexing (WDM) systems technology was rapidly evolving to offer easier solutions at lower speeds at the same time avoiding chromatic dispersion. The advances in WDM technology with a wide deployment also supported the Internet traffic needed at that time. With WDM deployment, the complexity of both electronic and photonic components shifted to a different paradigm.

As the telecom bubble started to rupture, business slowed down in some areas, with demand and sales volume smaller than expected. The first announced separations hit our lower technical staff, the first ones that I had trained and essential to lab work. Before one of the separation announcements, I realized I had more options than most of them. During my last year with special permission from JDSU, I had already been consulting to one of the DARPA programs under the MTO. As soon as I could, I became a full time technical consultant for Booz Allen Hamilton in Arlington, VA, but that was also for a short period.

8.4 Motivation to Join Academia

When I started graduate school in the USA, the population of women compared to all students among students seeking doctorate degrees in science and engineering was smaller or defined underrepresented. While the percentage of doctorate degrees in engineering (including computer sciences) and physical sciences earned by women may have increased since my graduation many women at the time of earning their doctorate degrees (like me) chose employment outside academia [20].

One motivation to be in academia is by having more faculty women professors and not teaching faculty, inside a classroom and being able to have a conversation with students about your career with your research and to relate teaching on a daily basis can make a difference. It is different from graduate students doing research for you. That would change the number of women studying engineering (or STEM disciplines in general) and effectively increase the number of women in academia representing the faculty equality as the men. The goal is to change their attitude and perception of women in general.

For me that was the major motivation to seek academic employment. As a part of being an academic faculty, I contribute to student's career, I even volunteer to go to recruiting for NC State. When I was a graduate student at Cornell, there was only one female assistant professor. She stayed briefly there. I can recall only male instructors; not any offerings of any engineering or physics courses taught by women instructors. For me to read about women scientists was history, different from my earlier graduate world. Although in Professor Eastman's group, there were several female graduate students.

When I joined NC State in 2003, I was the second tenured full professor female of the ECE Department after Dr. Sarah Rajala was hired in 1979.

8.5 Government Service: At National Science Foundation

The opportunity to serve as program director at the National Science Foundation was one of the greatest honors I had in my career. As IPA (the 1967 intergovernmental personnel act) assignee I served as program director for the Electrical, Cyber and Communication Systems Division for over 2 years, at the National Science Foundation (Engineering Directorate) in Arlington, VA. In addition to having the opportunity to start a new program initiative: the hybrid communications program. Prior to that, my experience with NSF had been only as a panelist for reviewing major instrumentation grants before academia.

The National Science Foundation is an agency that mainly funds research and education, structured like academia with disciplines unlike other federal agencies, it does not have a mission. To run their grants and programs the agency depends on 50% from borrowed personnel from academia or other agencies depends on directors (at least the time I served) through IPAs, who usually they start every summer. A natural renewal of staff comes from academia or other agencies to work there. The fresh brainpower is the engine that generates new ideas to run the different programs along with the expertise of the permanent staff. I learned a lot from government employees that interact with the "rotators" every day with knowledge and the know how to get it done. They were well informed, open to new ideas, and highly cooperative. NSF has a vibrant environment: new program initiatives that started at that time: on the Nanoscience, collaborate in other activities, and also donate my time for advice and mentor young faculty that just started their academic career. In the first few months I initiated a new solicitation for a hybrid communications program with small funding that was available for enabling technologies to sponsor research in free space optics, Terahertz components, RF and microwave devices, and subsystems.

During my brief tenure, there I was fortunate to interact with several talented individuals in the major cross cutting programs and the start of the Emerging Frontiers in Research and Innovation [21] program in 2007. At that time, the engineering program directors in a general workshop would brainstorm for collaborative multidisciplinary ideas. After ranking the most popular groups, the program directors would (some even belonged to more than one group) prepare a pitch to defend their topic to the administrative director for funding. In the first year, the topic of Resilient and Sustainable Infrastructures that I had contributed did not make the final two but on the second year made it!

The exposure and the service were invaluable, however, there were consequences at that time for being away so much from campus. Without a postdoctoral and small group, my research production had suffered. However, I had developed a wider network and more exposure, in addition to becoming an NSF personnel for life.

8.6 Back on Campus After NSF

Back full time on campus, I accepted to be the Director of Graduate Programs of the ECE Department for 1 year while the faculty that had served as such was going on sabbatical. I would be the first female Director of Graduate Programs (DGP) in history of the College of Engineering at NCSU. I did not know that until the Assistant Dean of the Graduate School told me. The ECE Department is one of the largest departments in the College and had one of the largest graduate programs among the Southeast Schools. At the time, the total number of graduate students was 650. The department offers two PhD degrees and several MS (with Thesis and non-thesis) in electrical engineering and computer engineering. Then only one MS non-thesis Program (Computer Network, shared with the Computer Science Department) had a dedicated advisor, the DGP (in this case, I) would advise all other programs. With the two staff members, and one in the whole university with an on-line web database for applications and review program with an excellent database that the former DGP had created and could help in the selection and tracking students. Later it was implemented widely to the college of engineering and then to the NC State Graduate School.

Following other models including from the computer sciences graduate program, I started a mandatory orientation course for all graduate students to provide the basic information on the department, graduate program, code of ethics, IT issues, how to navigate the library, and then an overview on intellectual property and overview of the faculty by personalized seminars. After the seminars I surveyed the students, the student cohort had and introduced them to each other and faculty, face-to-face meetings, and offering opportunities to meet at the same place every week. They would be able to create some subnetwork.

At the same time doing the graduate programs, I had also joined the group with a proposal for a new NSF-ERC on campus: Future Renewable Electric Energy Distributed and Manageable (FREEDM) Systems Center. One year later, I joined the leadership team of the new center as director for diversity and education that lasted for the next 3 years.

While doing my IPA at NSF, I started collaborating with one of my assistant professor colleagues as one of my graduate students was making targets for Indium Gallium Zinc Oxide, an amorphous semiconductor oxide for pulsed laser deposition system. These compounds have been considered as an alternative to amorphous silicon for display technologies and other applications such as sensors, electronic paper, and flexible electronics. When this faculty colleague also a Navy reserve officer recalled back to active duty in Iraq, my research effort increased on transparent devices based on amorphous Indium Gallium Zinc Oxide compounds as promising channel materials for thin film transistors into circuit design and integrated circuit realization [22].

8.7 Technical Volunteer and Citizenship

While at AT&T, there were several opportunities for participation in workshops and the general community. The shared knowledge either technical or personal was valuable because the community was open as I worked with friends, went to yoga or even played soccer with the brightest (I cannot play soccer!). It was the part of being in the community, however, the number of women researchers varied considerably on several parts of the organization. One particular organization that had more women was the materials and chemistry that included several department heads and had laboratory directors, while others had single digit numbers in whole laboratories. At least during the summer there was a considerable effort through the different fellowships and internships to bring a diverse number of undergraduate and graduate students. Some would come for one summer others would be able to return with fellowships or maybe apply for other positions. The network amplified in different magnitudes and contributed to the enrichment of everyone's life. It seemed natural to become a technical volunteer not only in serving in technical conferences or societies for reviewing manuscripts. There are about three organizations representing my technical fields of interest but IEEE covers all. Throughout my career, I intertwined volunteering with technical work because it is the way to payback what I learned and received from the conferences, meetings and publications.

8.8 Giving Back

I have served in different kinds of committees, working to improve the local environment and difficult to excuse myself. That includes being a faculty senator. I became involved in faculty policies and issues. With a very approachable vice chancellor for faculty affairs (Betsy Brown), and one of my senator colleagues following the initial research from the senate chair, we worked together to create an ombudsman position to faculty and staff to campus. While the topic had been around for almost 40 years, it never came through as solid.

8.9 Faculty Mentorship

New hires starting on tenure track can benefit from two kinds of mentors: one that understands the technical area and another who complements in the overall tenure/academic process. I have been a mentor with a process implemented in our ECE department and seen significant positive results. First, it helps the new faculty to establish and expand their own network, creating a personalized career path with guidance. Updating and having frequent contact between mentors and mentees create good and open relationships. It is also worth getting a mentor for grant writing, in addition to attending the offered workshops on campus.

Disclaimer As the author, I take responsibility for any errors of fact that appear within. I relied on my memory.

Acknowledgments I am deeply in debt to all teachers, professors, collaborators, managers, students, friends and colleagues that believed and invested in me. You know who you are! Your time spent with me and your words of advice were never wasted. I would like to thank my husband and my four children that make me a better person every day with their unconditional love.

References

1. Lunardi L (1975) Study of delayed neutrons from uranium and thorium photofission. https://bv.fapesp.br/pt/proc6291/14919//
2. Lunardi L (1979) Electrodesintegration of the ^{232}TH by emission of a neutron. https://bv.fapesp.br/pt/proc6291/15161//
3. Miller SE, Kaminow IP (1988) Optical fiber telecommunications. Academic Press, San Diego, CA. ISBN: 0124973515, 9780124973510
4. Malik RJ, Lunardi LM, Walker JF, Ryan RW (1988) A planar-doped 2D-hole gas base AlGaAs/GaAs heterojunction bipolar transistor grown by molecular beam epitaxy. IEEE Electron Device Lett 9(1). https://doi.org/10.1109/55.20396
5. Swartz RG (1987) Unpublished results
6. Swartz RG, Lunardi LM, Malik RJ, Archer VD, Feuer MD, Walker JF, Fullowan TR (1989) AlGaAs/GaAs heterojunction bipolar transistor decision circuit. Electron Lett 25(2). https://doi.org/10.1049/el:19890087
7. O'Connor P, Flahive PG, Clemetson WJ, et al (1984) Monolithic multigigabit/sec GaAs decision circuit for lightwave system applications. In: conference on optical fiber communication location Digest of Technical Papers Pages, pp 26–27, New Orleans, LA, 23–25 Jan 1984
8. Ohta N, Takada T (1983) High-speed GaAs SCFL monolithic integrated decision circuit for Gbit/s optical repeaters. Electron Lett 19(23). https://doi.org/10.1049/el:19830668
9. Chandrasekhar S, Campbell JC, Dentai AG, Joyner CH, Qua GJ, Gnauck AH, Feuer MD (1988) Integrated InP/GaInAs heterojunction bipolar photoreceiver. Electron Lett 24(23). https://doi.org/10.1049/el:19880986
10. Chandrasekhar S, Lunardi LM, Gnauck AH, Ritter D, Hamm RA, Panish MB, Qua GJ (1992) A 10 Gbit/s OEIC photoreceiver using InP/InGaAs heterojunction bipolar transistors. Electron Lett 28(5):466–468. https://doi.org/10.1049/el:19920294
11. Lunardi LM, Chandrasekhar S, Gnauck AH, Burrus CA (1995) 20-Gb/s monolithic p-i-n/HBT photoreceiver module for 1.55-μm applications. IEEE Photon Technol Lett 7(10). https://doi.org/10.1109/68.466590
12. Reichmann K, et al (1996) Four-channel p-i-n/HBT monolithic receiver array at 500Mb/s per channel for local access applications. unpublished results
13. Rios JMM, Lunardi LM, Chandrasekhar S, Miyamoto Y (1997) A self-consistent method for complete small-signal parameter extraction of InP-based heterojunction bipolar transistors (HBT's). IEEE Trans Microw Theory Tech 45(1). https://doi.org/10.1109/22.552030
14. Mehta JU, Borders WA, Liu H, Pandey R, Datta S, Lunardi L (2016) III–V tunnel FET model with closed-form analytical solution. IEEE Trans Electron Devices 63(5). https://doi.org/10.1109/TED.2015.2471808
15. Chandrasekhar S, Lunardi LM (1995) High-performance p-i-n/HBT monolithic photoreceivers for lightwave communications. In: Koteles ES; Willner AE (eds) Emerging components and technologies for all-optical networks. AE Book Series: Proceedings of the Society of Phot-

Optical Instrumentation Engineers (SPIE), vol 2613, pp 91–97. Proceedings paper: conference on emerging components and technologies for all-optical networks, Philadelphia, PA, Oct 24, 1995
16. Garrett LD, Chandrasekhar S, Zyskind JL, Sulhoff JW, Dentai AG, Burrus CA, Lunardi LM, Derosier RM (1997) Performance of eight-channel OEIC p-i-n/HBT receiver array in 8x2.5 Gb/s WDM transmission system. J Lightwave Technol 15(5):827–832. https://doi.org/10.1109/50.580823
17. Ota Y, Swartz RG, Archer VD (1992) DC-1 Gb/s burst-mode compatible receiver for optical bus applications. J Lightwave Technol 10(2):244–249. https://doi.org/10.1109/50.120581
18. Lunardi L, Chandrasekhar S, Swartz RG, Hamm RA, Qua GJ (1994) A high speed burst mode optoelectronic integrated circuit photoreceiver using InP/InGaAs HBT's. IEEE Photon Technol Lett 6(7). https://doi.org/10.1109/68.311464
19. NSF-STATS. https://www.nsf.gov/statistics/tables-by-survey.cfm
20. NSF-EFRI Emerging Frontiers in Research and Innovation (EFRI-2008), Resilient and Sustainable Infrastructures (RESIN). https://www.nsf.gov/pubs/2007/nsf07579/nsf07579.htm
21. Luo H, Wellenius P, Lunardi L, Muth JF (2012) Transparent IGZO-based logic gates. IEEE Electron Device Lett 33(5). https://doi.org/10.1109/LED.2012.2186784
22. Chandrasekhar S, Zirngibl M, Dentai AG, Joyner CH, Storz F, Burrus CA, Lunardi LM (1995) Monolithic eight-wavelength demultiplexed receiver for dense WDM applications. IEEE Photon Technol Lett 7(11):1342–1344. https://doi.org/10.1109/68.473492

Chapter 9
Integrated Circuits, MEMS, and Nanoelectronics for Sensor Applications

Mona E. Zaghloul

9.1 Background and Education

I was born in Cairo, Egypt at a time during which Egypt was still ruled by the Royal family. My family was an upper middle-class family. Both of my parents were highly educated, and both had obtained their PhD degrees in chemistry from England. Both were Professors of Chemistry at Cairo University, which at the time was a highly prestigious university. My mother and father spent a considerable amount of time in their laboratory at the university, and they often took me with them. There I watched them work in their labs, mixing chemicals and writing notes. I watched the long hours of hard work they both spent while trying also to be good parents, always taking care of me, and providing me with things to keep me busy.

My father passed away when I was 11 years old, and my mother subsequently had to take care of me and my sister alone. I finished high school with above average grades in 1960. My mother insisted that I pursue an engineering education (my obvious wish was to be a chemist) and that I apply to the engineering school at Cairo University. Cairo University was one of the top engineering schools in Egypt and accepted only the most accomplished students with the highest grades. I was accepted and chose to study electronics there. We were educated by Professors who received their PhD degrees from mostly European countries, the most common being the United Kingdom. We had a good education at the time because the laboratories, and the university system as a whole, were set up to follow the British System. By the end of my undergraduate career, I met my husband who was a classmate, and we were married soon after graduating.

M. E. Zaghloul (✉)
Department of Electrical and Computer Engineering, The George Washington University, Washington, DC, USA
e-mail: zaghloul@gwu.edu; http://mems.seas.gwu.edu

A. C. Parker, L. Lunardi (eds.), *Women in Microelectronics*, Women in Engineering and Science, https://doi.org/10.1007/978-3-030-46377-9_9

My husband was awarded a fellowship to pursue his PhD in Electrical Engineering at the University of Waterloo in Ontario, Canada. I traveled with my husband to Canada in 1968 and joined the Department of Electrical Engineering at the University of Waterloo as a Research Assistant. I continued my studies there and obtained my master's degree in Electrical Engineering in 1970. After my master's degree, I joined the School of Mathematics at the University of Waterloo to pursue a second master's degree, this one in Applied Mathematics and Computer Science. At that time, the University of Waterloo was a pioneer in the study of computer science and programming and was known for its programming language WATFOR. I graduated with my second master's degree in 1971.

Towards the end of my degree, I was offered an opportunity to pursue a PhD with a professor of Electrical Engineering whose work focused on computer aided design of nonlinear electronics and circuits. At the time, many schools were trying to develop computer simulations for active transistor circuits. CAD tools such as SPICE were first introduced around that time for simulating and designing circuits. My thesis work focused on modeling circuits and electronics.

The time that I spent working on my PhD in the early 1970s was a unique and exciting time. At the time, Waterloo had a computer network that connected the entire university campus. This network was used by faculty and graduate students to send emails and to transfer data. This was the early Internet, and our network, called WITS, was one of the first of such networks in the world. In addition, the members of our department had many close contacts and collaborations with members of the Department of Electrical Engineering at University of California, Berkeley. We often had visitors and professors from Berkeley spend a semester or more at Waterloo, teaching graduate students and using Waterloo computer facilities for their own work. I learned a lot from these interactions during my PhD work.

I was awarded my PhD in 1975. I was the first woman to graduate with a PhD from the School of Engineering at the University of Waterloo, Canada. I did not know this at the time. However, in retrospect I had noticed that I was often the only woman in attendance in most of my engineering courses. This was very different from my undergraduate experience. Although women still only represented a minority of engineering students at Cairo University, my class at Cairo nonetheless still had nine female students (compared to 120 male students). I believe part of the reason behind why there were more women studying engineering in Cairo was because the country had just been through a revolution, and at the time, there was a great desire to build the country. The country needed engineers, and I remember several advertisements designed to recruit all kinds of engineers. Many years after I was awarded my PhD degree in 2007, the University of Waterloo awarded me an honorary PhD degree in recognition of the fact that I was the first female graduate from their School of Engineering, and for my career and my accomplishments in research and teaching.

9.2 Move to the USA and Early Career

We moved to the USA in 1978. My husband was offered a position at COMSAT, Inc., a communications satellite company in the Washington DC area. At the time of our move, I had two young children—a 4-year-old boy and a 4-week-old girl. I moved to the DC area with my family and started looking for a job.

I was interested in working in academia because I had seen my mother and the example she had set. It was difficult, though, largely because I was a woman. I did not know at the time that female engineers were rare in the USA. I met an individual early in my job search who told me simply, "Ma'am, you are a woman, you are an engineer, and you are a foreigner (thinking I was French Canadian). You think you will find a job in this place? This is Washington, DC, the top of the nation." Many of the recruiters I met told me that it was near impossible for a woman with a PhD in engineering to find a job. My job search was more difficult, of course, because I had a husband and two small children, and was limited in terms of the geographic area in which I could search for and accept a position.

I was eventually offered a job at the Computer Science Corporation, mostly because the Department of Computer Science at the University of Waterloo had an excellent reputation and was known for its programming language WATFOR. My first job involved programming using several of the programming languages I had studied and drawing upon the experiences that I had writing code at Waterloo. Although my PhD was focused on modeling circuits, I accepted this initial position to work as a computer scientist while I continued my search for an academic job in my major area of expertise. I continued writing papers on my PhD work during my free time and submitting them for publication. It was difficult, but after several attempts, I was finally offered a job as an Assistant Professor in the Department of Electrical Engineering and Computer Science (EECS) at the George Washington University in Washington, DC.

I was the first woman to be hired in the School of Engineering and Applied Science at GWU. This was a big change for many other members of the faculty, and many did not accept this change. The Chair of the EECS Department at the time was Professor Ray Pickholtz. He was well known in the field of communications, and he certainly believed in my ability to work and to produce quality research. In addition, he knew that I could teach computer science courses in addition to courses in electrical engineering. His support during this early time was critical.

I was offered the job at The George Washington University because of my knowledge of computer science and because of my extensive programming experience. I had experience with developing computer aided design tools (CAD tools) for electrical circuits, and with developing simulations of circuits with active components. In addition, I was very familiar with circuit theory as result of the courses I attended during my PhD graduate courses that were mainly taught by Berkeley professors such as Professor Leon Chua. Dr. Leon Chua and his postdoctoral fellows taught us circuit theory including the MEMRISTOR device, and its modeling, and its circuit implementation.

I started my teaching position at GWU in 1980 by introducing several new courses on circuit theory, computer aided tools for circuit simulation, and on modeling linear and nonlinear circuits. I also taught courses on optimization, graph theory, and programming in different languages. These changes were unique to the graduate curriculum at GW and the courses attracted a number of students. At the time, my courses were introducing new knowledge in computer science, and the students were appreciative. My students were mainly from government agencies around the DC area who were attending evening classes.

In 1984, I applied to a position at the National Institute of Standards and Technology (NIST), which was called the National Bureau of Standards (NBS) at the time. I was hired by Dr. Ken Galloway, the head of the Semiconductor Electronic Technology Division. I applied for the position because I was interested in pursuing research and was trying to find better facilities than were available at GW and because I was trying to create collaborations in the area. I was hired as a "faculty hire," to go to NIST 1 day a week in addition to my teaching job at GWU. The NIST job was an important step for my career. I was exposed to the best laboratories and the best research scientists specializing in the area of semiconductor electronics research. I was exposed to extensive techniques for measuring and testing electronic circuits and for testing electronic materials and applying the standards of the industry.

In 1984, a course was offered by MOSIS (MOS Integrated Systems was funded by the US Government) for learning the design of integrated circuit chips. The course was offered by top leading university professors at the time and the goal of the course was to teach other university professors, and consequently their students, the art of VLSI chip design. Professor Carver Mead from California Institute of Technology was the lead Professor. The course was offered at the MOSIS building at Marina De Ray for 2 weeks. This was a new initiative set up by the US government to motivate academic researchers to begin designing and building integrated circuit chips. I learned many things from this course. I started designing complementary metal oxide silicon (CMOS) chips for fabrication and testing. This was a skill that was very much in demand in the Semiconductor Electronics Technology Division at NIST where I worked. It was important to design test structures to measure the performance of devices for different technologies. I collaborated with many of my colleagues in the division on these tasks. In addition, I started teaching chip design at GWU and started sending our own chips to MOSIS for fabrication. We sent several students' projects using analog and digital integrated circuit designs to MOSIS for fabrication. MOSIS returned the fabricated chips to GWU where we tested the chips and prepared testing reports for submission back to MOSIS.

I added several other courses to the GWU curriculum to reflect the depth of CMOS design. I wanted to carry several research projects and at various applications using CADENCE software. I started collaboration with University of Maryland Professor Robert Newcomb on designing Neural Networks and trying to implement the neuron with the ability to learn and processing data. Several students were trained in the topic of building hardware for learning neural network and as a result, several students earned their PhD in this topic. As a consequence, several GWU students graduated with the knowledge of CMOS chip design and with my NIST knowledge of the industry standard and requirements that was reflected in their courses. Many of these students are now part of a productive force in top IC design companies such as Intel, Apple, and others. Because of our knowledge of the theory and design of Neural Networks, we learned about its applications to the classification of data and especially for large data (Big Data). Thus, I was asked by my NIST colleagues to develop a machine learning approach for classifying large semiconductor data, and I was asked to write the code and verify it (that was in 1994).

In 1996, I was awarded IEEE Fellow for my work on Integrated Circuits chips design and for the Circuits we developed for implementing Neural Networks mimicking the firing and the learning of the neuron. In 2017, I was selected as a Fellow of the National Academy of Inventors (NAI) recognizing my patents in the areas of microelectronics and MEMS.

In 1990, the NIST team started paying attention to MicroElectroMechanical Systems (MEMS) design and expanding the scope of the CMOS materials to all possible other material. I joined this effort at that time. In the MEMS work, you have to be very familiar with the different layers of the material. The designer has to add and to subtract new materials for the construction of MEMS device. The designer must have extensive knowledge of deposition and etching of different materials, knowledge of material interaction, and extensive knowledge of clean room facility equipment, knowledge of the use of imaging equipment to verify the designed and fabricated devices. I had a large team of students from GWU to help me, to try several recipes, and to work on developing novel devices for MEMS applications. The applications chosen were to develop sensors. We used most of the available CAD tools for us to simulate the MEMS devices. Our team designed several novel MEMS devices. The devices were patented and were acquired by industry. I had several PhD students working with me at NIST. We used MEMS to implement varieties of novel sensors and it was an opportunity to add the electronics on the chip to yield smart sensors. I also used CMOS technology to build novel MEMS devices by adding to the CMOS technology other materials on the surface of the fabricated CMOS device (Surface Micromachining), or by etching layers from the standard CMOS technology (Bulk Micromachining) or possibly use both techniques at the same time. We developed several novel devices and sensors with integrated electronics. An example was developing mechanical Surface Acoustic Wave (SAW) device for resonators applications and for biosensors.

The work was acknowledged, and several students graduated using these techniques and many joined the industry for MEMS design and with the knowledge of IC design, they were hired as VLSI designers in addition to the skills of MEMS designer. Recently, I joined material science group at NIST to synthesize 2D materials and start designing electronics circuits using 2D material. I have been working with Material synthesis group at NIST for more than 5 years on 2D material. My students and I designed different transistors using different 2D materials and using 2D materials for sensors. For the design of 2D materials, the designers have to know how to build the devices in the clean room and be able to measure the devices and using imaging equipment to have an image of the device. Several devices were fabricated and tested at NIST laboratory and at my GWU laboratory. Recently, GWU added a clean room to its new facility, which allowed our students to use the new facility for building novel devices and using excellent imaging equipment to characterize the fabricated devices.

In 1989, I was promoted to Full Professor at the Department of Electrical and Computer Science. Again, I was the first woman to be appointed Full Professor at the George Washington University School of Engineering and Applied Science (Fig. 9.1).

This was big step and I think it opened the door for many other women to follow. At that time, several female professors were hired at GWU School of Engineering, and a different whole era started.

The number of women engineers is still small in proportion to all engineers. For example, in the USA and Canada, it is about ~11%, and in Australia it is about 9.6%.

To the best of my knowledge other European countries such as Latvia, Bulgaria, Cyprus, and Sweden have 30%, 29.3%, 28.6%, and 25.9%, respectively. The percent of women engineers in China is 40%.

The IEEE, Institute of Electrical and Electronics Engineers, International organization, headquartered in New Jersey, USA created Women in Engineering Organization (WIE) as an affinity group in 1994. WIE celebrated its twenty-third anniversary in the year 2017; estimated number of members is 24,000.

WIE is very active in many regions in the USA and globally: Asia, Latin America, Europe, and Africa.

Fig. 9.1 Zaghloul with her students in the Research Laboratory

In Spring 2018 the GWU School of Engineering and Applied science undergraduate classes 2018 is about 25% female students, which is considered one of highest in the country. Thus, several female undergraduate students are attracted to Engineering, and the female engineers are now recognized and respected as productive engineers.

I teach several female students in my classes and I am proud of their accomplishments. Many professional societies recognize women in engineering and encourage women to join the profession. I am impressed by the progress of female engineers since I started my career, and certainly, it was long road since 1970.

Examples of Devices Developed at Zaghloul Laboratory

Professor Zaghloul works with her student in her research laboratory as shown in Fig. 9.1. In the following, I will briefly give examples of devices developed in our laboratory. Many of the devices were micro/nano-fabricated in clean room, and mostly are patented (Figs. 9.2 and 9.3).

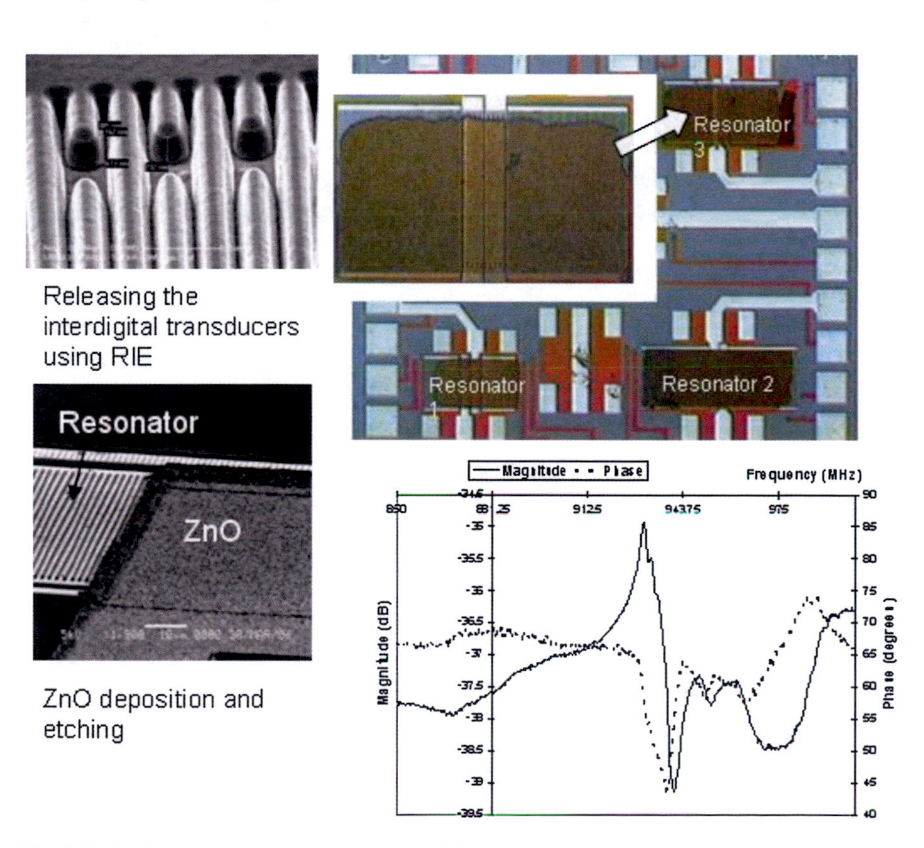

Fig. 9.2 Surface acoustic wave resonators, which are used as filter for transmitter/receiver systems. It was developed using CMOS technology with post-processing steps to obtain the resonators. Reference: *Anis N. Nordin, Mona E. Zaghloul, "Modeling and fabrication of CMOS Surface Acoustic Wave Resonators"; IEEE Transactions on Microwave Theory and Techniques, Vol. 55, No.5, May 2007, pp. 992–1002*

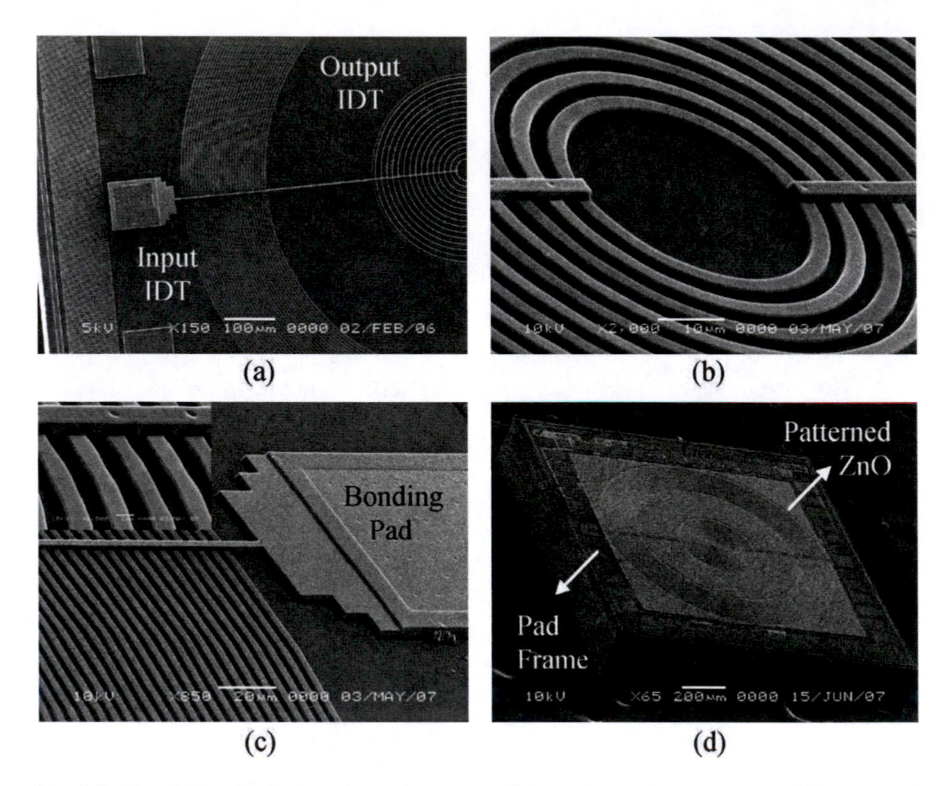

Fig. 9.3 Novel Circular Surface Acoustic wave, which can be used as resonator or biosensor. (**a**) Top view scanning electron microscope (SEM) of the circular SAW chip after CMOS and before post-processing. (**b**) Close-up view of the inner IDT after RIE step with 90 sidewalls. (**c**) Close-up view of the outer IDT with the pad after RIE. (**d**) Top view SEM snapshot of the device after final step of ZnO patterning. Reference: *Onur Tigli, Mona E. Zaghloul; "Design, Modeling, and Characterization of a Novel Circular Surface Acoustic Wave Devices"; IEEE Sensors Journal, Vol.8, Mo. 11, November 2008, pp 1807, 1815*

Recently, my research included nanoelectronics material for implementations of nanotransistors (Fig. 9.4).

Measurement of the transistor characteristics is shown in Fig. 9.5.

Currently, we are working on Health Care Sensors. Precision health care is shown in the following Fig. 9.6, where sensors implanted in the body or attached as wearable or distributed in smart home generate the health data. The data are analyzed, and action response is determined.

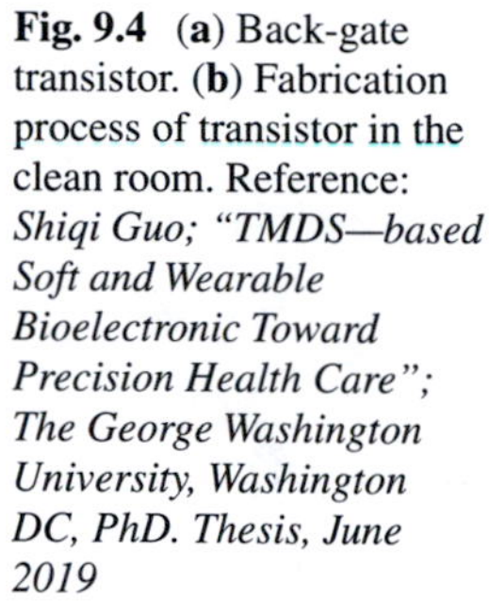

Fig. 9.4 (**a**) Back-gate transistor. (**b**) Fabrication process of transistor in the clean room. Reference: *Shiqi Guo; "TMDS—based Soft and Wearable Bioelectronic Toward Precision Health Care"; The George Washington University, Washington DC, PhD. Thesis, June 2019*

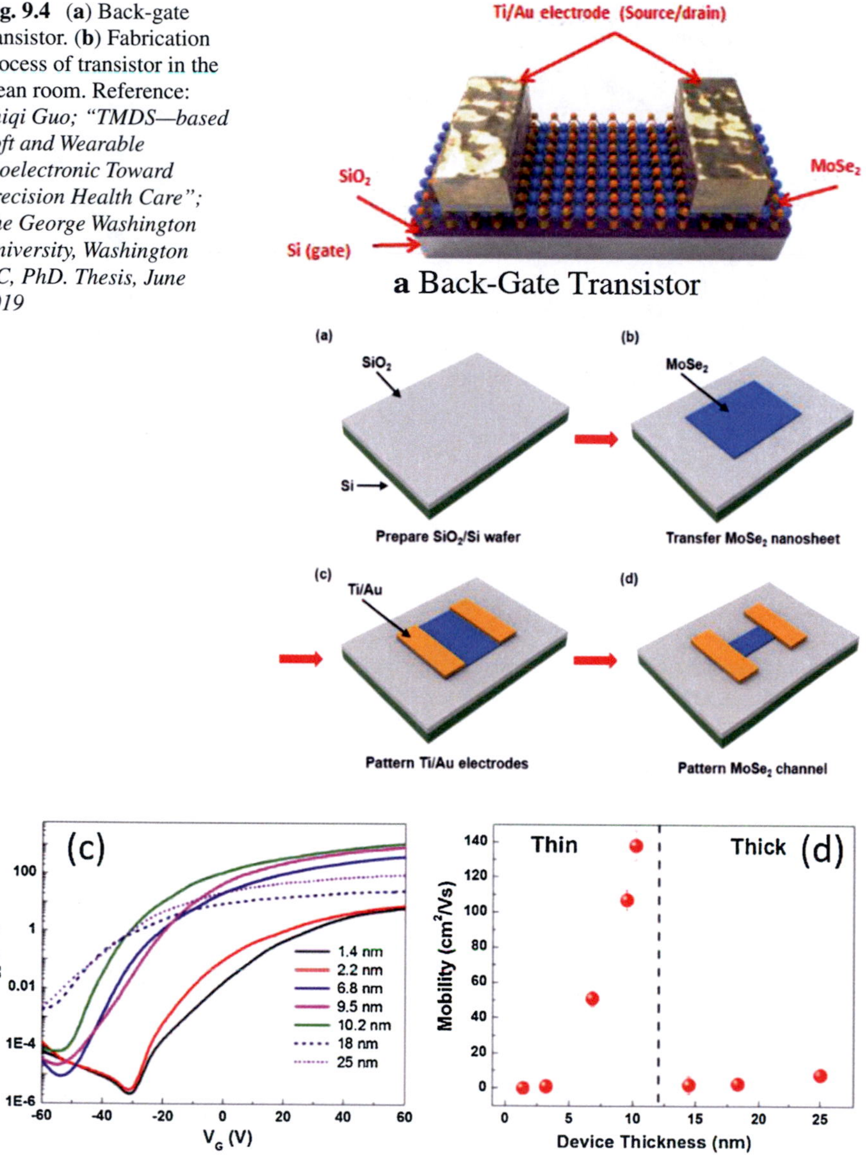

Fig. 9.5 Measurements characteristics of nanotransistor. Reference: *Shiqi Guo; "TMDS—based Soft and Wearable Bioelectronic Toward Precision Health Care"; The George Washington University, Washington DC, PhD. Thesis, July 2019*

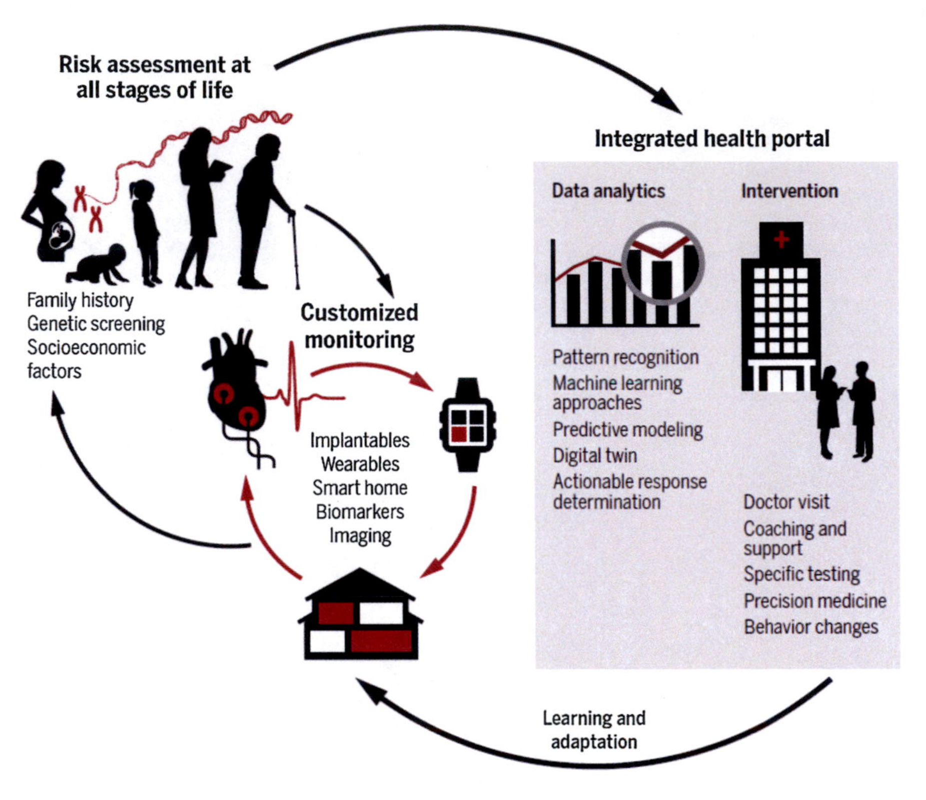

Fig. 9.6 Precision health care overview. (*Sci. Transl. Med.* 10, eaao3612 (2018))

9.3 Professor Experience at the George Washington University: Challenges and Successes

In the early years at GWU, I was repeatedly presented with the most challenging assignments. It seemed to me that my fellow male colleagues were eager to see whether a woman could really do what was, at the time, still primarily a man's job. My only choice was to respond to their increasing challenges with increasingly hard work, and with the tools, my education provided me.

In 1994, I was elected the Chair of the Electrical and Computer Science Department.

I was the *First Woman* to be Department Chair in the School of Engineering and Applied Science. The EECS department was large department with the largest number of faculties in the whole university.

I wanted to keep my research work going because I really like to do research. I succeeded in managing the department and we received accreditations from ABET with no problems for 6 years. I hired several assistant professors and associate professors and tried to increase the areas of research of the department. I kept my work with NIST and was at NIST with my students learning several new technologies and new directions, producing new devices and educated my students in the new directions of technology. I was the Chair of the Electrical Engineering and Computer Science Department for the period of 1994–1998. I ended my first period as chair. I went for sabbatical at Delft University in the Netherlands.

In 2000, the EECS department was separated into two departments, the Electrical and Computer Engineering Department and Computer Science Department.

In 2009, I was elected by my colleagues as Department Chair for the Department of Electrical and Computer Engineering. I worked as Chair from 2009 to 2014. I kept my research, kept my contacts with NIST, and was always eager to learn new technology and to introduce new technology courses to my students. Thus, I spend the time to learn nanotechnology and specifically nanoelectronics. I taught new course on nanoelectronics and introduced the students to nanofabrication and now with the new clean room at GWU the course is taught with the clean room laboratory to teach the students nanofabrication.

Currently, I am working with Material scientists at NIST to synthesize 2D material, and we are using 2D materials for electronics applications as flexible electronics. 2D materials are atom thick layers of materials, and it becomes active research areas after the discovery of graphene.

At 2014, I was hired as Program Director at National Science Foundation at the Division of Electrical, Communication, Cyber, and Systems (ECCS). I was responsible for Circuits, Communications, Sensors, Systems (CCSS) program. I worked at NSF from January 2014 to December 2016. It was very productive time and certainly, I learned and met many new researchers in my areas of interest. In 2017, I returned to The George Washington University and concentrated on my research.

Currently, I am concentrating on the following areas of research

- Research on integrated circuit design, as interface for sensors, complete with fabricating, and testing chips (mostly CMOS technology) with emphasis on flexible electronics components and integrating small CMOS components on flexible substrate.
- Research on MEMS/NEMS design and fabrication to develop novel sensors devices using different materials with emphasis on Biosensors for health care systems.
- Program on nanotechnology with emphasis on realizing nano-devices for electronics and sensors applications.

We are currently working on several projects under the above topics.

I promoted women in engineering and encouraged women undergraduates to be active in research. Over my career of 39 years at GWU, I graduated 41 PhD. The total number of females of all the PhD. I graduated are 8. That is about 20% of females completed dissertation under my supervision, and for 24 Master thesis under my supervision only 4 were females around 16%.

All female and male graduates are currently working in industry (Silicon Valley, etc.) and in academia (USA and abroad), with several having leading positions.

At the GWU School of Engineering and Applied science SEAS, the number of undergraduate female students improved at a considerable rate, reaching 39.8% of total number of students in 2018. For *graduate students*, the percentage of female students reached 26.7% of total students, which is above average nationwide. While this certainly leaves plenty of opportunity for growth, we must recognize and take pride in the gap that we have narrowed in the past 40 years.

9.4 Future and Prospective

I introduced several new technical initiatives, several courses, and research areas during my career:

- Program on integrated circuit design, complete with fabricating, and testing chips (mostly CMOS technology).
- Program and Institute on MEMS/NEMS design and fabrication.
- Program on nanotechnology with emphasis on realizing nano-devices for electronics and sensors *applications*.

There are many directions of research with the emphasis on health care, and it is true that in the near future we will have sensors attached or embedded in our body for complete monitoring of our health care. We are living in a very exciting time. New technologies are being developed around us all the time. We see tiny machines the size of one billionth of a meter. We are building with the individual atoms. This is an era of melding between man and machine. We are at a multidisciplinary time where the boundaries between electrical, mechanical, civil, chemical, and bioengineering are rapidly fading. *This is truly the era of science and engineering* and it is an exciting time to be an engineer or a scientist. The possibilities to create novel things appear limitless.

For the young women who wants to study engineering the opportunities in this field are limitless. There are many directions and being engineer is an outstanding opportunity. You yourself are the only limit to your achievements; with hard work and persistence, you can achieve whatever you set your mind to. It may not have been done before, but that does not mean that it cannot be done. Education is important and it gives you the foundation and the confidence that help shape your career.

Chapter 10
From Microwave Communication Systems to Nanomedicine Tools: Using Advanced Microelectronics Fabrication as an Enabler

Rhonda R. Franklin

10.1 Serendipity

10.1.1 My Family

I grew up in Houston, Texas with my parents, Ann and Elvin Franklin Jr. Both families valued education, faith, and love but were poor. My grandparents had at most an elementary school education. My mom's mother dreamed of a college education and her dad got his GED at the age of 58. My parents met at Grambling State University in Grambling, Louisiana, a historically black college and university during segregation in the south. With hard work and determination, they became first-generation college graduates and teachers in math (mom) and industrial arts (dad). These achievements prepared my parents for new opportunities.

My brother Fred and I were born just after desegregation and civil rights laws passed. These new laws with a college education would present many new doors of opportunity to my parents. It took courage and bravery for them to leave their families and communities to make a better life for us. After several moves, we landed in Houston, Texas when I was four. My mother taught remedial math in junior high school and my father became a manager and then a small business owner of an insurance agency.

R. R. Franklin (✉)
University of Minnesota, Twin Cities, Minneapolis, MN, USA
e-mail: rfrank01@umn.edu

A. C. Parker, L. Lunardi (eds.), *Women in Microelectronics*, Women in Engineering and Science, https://doi.org/10.1007/978-3-030-46377-9_10

10.1.2 It Is the Little Things

I had very few experiences before college to shape my interest in electrical engineering. The first two were subtle. I am told at four, I stuck a hairpin in an outlet and the current threw me across the room. At eleven, I warmed food in a paper container in our microwave oven. I did not notice the metal handle until our microwave was on fire. The last one was profound. I attended a summer engineering camp before my senior year. It influenced me so positively that I decided to become an engineer instead of the doctor, lawyer or musician I was contemplating.

10.1.3 Environmental Influences

The NASA space program and its images of the planets, stars, and galaxies were everywhere in Houston. Those images strongly influenced my interest in space exploration. I loved to read science fiction, solve problems, and understand how things worked. It would take decades to connect this interest to my chosen career in microwave engineering.

10.1.4 Childhood Experiences

My journey to becoming a microwave engineer and professor started with several childhood activities in the arts, a left-brain activity, and math and science, a right-brain activity.

10.1.4.1 The Arts: Dance

I trained for 11 years with Mrs. Delice Teamer to become a classical ballet dancer. I was tall and quick thinking, but inflexible and flatfooted. She taught me how to manage and control my body, how to be patient with myself, and how to find my body's voice and style. By eighteen, I was a graceful and elegant dancer. *She knew I was destined for college and was not going to be a dancer, yet, gave me her best and taught me the power of encouragement, self-discipline, and striving for excellence.*

10.1.4.2 The Arts: Music

I learned to play the trumpet in sixth grade from my first band director, Mr. Algernon Jones, a professional jazz trombone player. My connection with the instrument and music resulted in my practicing morning, noon, and night at home, before, and after school with him. My quick improvement resulted in an invitation to play with the eighth grade advanced band in my first year. It offered more challenging music and opportunities to learn from more experienced players. He also recommended getting a tutor, going to band camps and playing in the Houston Youth Symphony. I did them all. By eighth grade, I was the lead trumpet and loved classical and symphonic music. *He showed me the power of practice, having strong fundamentals, challenging one's self, and learning from people who care, believe in you and are better than you.*

Mr. James P. Moseley, my second band director in high school taught me how to push through inner challenges. I became section leader of the marching, concert, and jazz bands by my second year. My love of playing music was challenged by bouts of stage fright due to anxiety during solo performances. *He encouraged me to push myself through it and provided more opportunities to overcome stage fright.*

10.1.4.3 STEM: Math and Science

I always loved math with its symbols, and how they could be used to manipulate or represent different things. My earliest memory of math is with my grandfather, who patiently explained problems from his math books when we visited. At home, I played with sample math textbooks from mom's school.

I learned math and science from excellent, devoted, and supportive teachers who challenged me and were my role models. I was one year ahead in math each year starting in sixth grade. I took the next course until eighth grade when my teachers prepared special assignments for the next level of math. In high school I took the all available math classes also (e.g., Algebra I and II, Geometry, and Trigonometry). My teacher, Mr. Ali Saleem (aka Mr. Hines), who was working on a PhD in math topology, developed a pre-calculus course for me so I could be better prepared to study engineering in college based on the summer engineering program director's recommendation.

For a historical context, between 1973 and 1983, my neighborhood changed quickly. My predominantly white elementary school transitioned very quickly to African American starting in middle school due to white flight. Most of my teachers starting in fourth grade were African American, including in math, science, and music.

10.1.5 The Power of Outreach

My physics teacher, Mr. Dupas, recommended I attend a National Science Foundation summer engineering program (MESET) at the University of Houston. The Director, Dr. Paskusz, developed and led an excellent program that provided guidance, skill building, peer tutoring, mentoring, community building, and career presentations by faculty. We spoke recently for the first time. He is in his nineties.

These different experiences were very important to my becoming an engineer. The arts developed my character and taught me self-expression and pushing through challenges while the STEM experiences piqued my interest and led me to a career path for those interests. When asked, how did I choose electrical engineering? My authentic answer is unexpected. I did not know any engineers, but I knew I wanted to explore being an engineer. To break a tie between mechanical and electrical engineering, I let my fascination with the letter "L" be the tiebreaker. It was the right decision for me.

10.2 On Becoming an Engineer

My first three years at Texas A&M University in College Station, Texas were a mix of learning engineering basics, building community and developing leadership skills.

College instruction did not provide content connections or context. I also missed the one-on-one discussions for meaning and understanding. Math was hard, yet interesting, and doable. Second term physics was not so doable. I almost changed my major after the electricity course but decided to finish the term and give myself one more semester to see if engineering could be a good fit. Glad I did. The third year of physics combined basic circuit courses and good instructors were awesome! I felt I was on the right track and never looked back.

One challenge was dealing with the nagging feeling of isolation in engineering and on-campus. My demographic was less than 0.3 % of the total campus population at the time (36,000+ in 1983). Therefore, I was involved with different campus activities to create community with women and African American students (e.g., Memorial Student Center student counselor, black student union, Alpha Kappa Alpha sorority, National Society of Black Engineering society, etc.) and to develop soft leadership and organizational skills outside of engineering. When the College of Engineering asked me to carry the college flag at graduation as a representative of the college and my peers, I was honored. It made me feel appreciated, accepted, and valued in the college.

I sought internships to understand what I could do with an electrical engineering degree. As a first-generation engineer, I was unclear about specialty areas for technical electives. Summer internships were invaluable for providing exposure. I tried practical and research internships. I preferred research ones that challenged me to use and grow my technical skills. Dr. Mark Eshani mentored me for two years and encouraged me to consider power electronics. After a course on power electronics my junior year, I wrote my first research interest statement for an internship at Lawrence Livermore National Laboratory (LLNL) with his help. *This opened the first door to my finding microwaves and microelectronics.*

The LLNL research internship involved high voltage (1000s of volts) power electronics and required constant supervision. In addition to reading, I observed testing by technicians, and having the technicians spot me while I performed tests. Their joking and teasing about a variety of things, however, made me feel uncomfortable and worried about my safety.

Dr. Paul Phelps, an engineering leader, asked about my summer experience. My response of "okay," prompted him to ask what would make it better. I responded: "more autonomy." He suggested working at microwave frequencies where parts would be small and low power. His knowledge of Livermore projects impressed me and prompted me to ask, "How can I do what you do someday?" He said, "Get a Ph.D." Therefore, that day, I decided I was getting a PhD and took my technical electives on ultra-high frequency circuits. *My journey with microwaves began.*

10.2.1 Planning for Graduate School

Dr. Ehsani was understanding of the change and wrote positive recommendation letters that helped me get into the various graduate schools. Dr. Karan Watson also supported me with recommendation letters. Dr. Howard Adams, the Executive Director of the GEM Fellowship program, promoted graduate education and the GEM MS fellowship program annually on-campus. With its consortium of companies and university partners, LLNL selected me and became my sponsor. I was accepted into all of the recommended microwave programs and chose to attend Michigan after the positive interactions with Dr. Linda Katehi about her research and the program. Feedback from trusted other GEM fellows was reassuring of the decision. This decision required me to be courageous and brave like my parents and leave my supportive home and community to become a microwave engineer at Michigan. *This was another life changing decision.*

10.3 Graduate School

10.3.1 Microwave Space Communications and Microelectronics

Graduate school re-connected me to space applications. I had classes from leaders in the field of microwave applications. I worked with Dr. Linda Katehi, an expert in microwave circuits, antennas, and modeling in the Radiation Laboratory.

Drs. Karan Watson and Linda Katehi would be the only female electrical engineering professors I would have in my academic journey as a student. These exceptional researchers and leaders were also supportive and encouraging mentors.

10.3.2 Timing Is Everything

It was a great time to be a graduate student in the Radiation Laboratory from 1988 to 1995. NASA's interest in "establishing operational bases on the moon, manned and unmanned operations to Mars, and space flight missions to other parts of the solar system, and observations of the Earth's environment from space platforms [1]" created many research opportunities. To do so, performing these experiments from space to overcome the limitations of significant signal attenuation and noise from earth to space was deemed essential.

Michigan's NASA funded Center for Space Terahertz Technology program in 1988 aimed to research key technologies and build systems using terahertz devices, circuit, antenna arrays, and heterodyne receiver systems. These systems would support molecular spectroscopy of planetary and the Earth's upper atmospheric layer. Research focused on developing device modeling, fabrication, and performance evaluation initially in the 0.1–1 THz range. It would also design solid-state receivers that included antennas and integration of a mixer and oscillator. These designs would eventually be used to study the Earth's upper atmosphere, investigate astrophysical events, and pursue industrial applications [1].

The lack of good sources—diodes and/or transistors, double-ended balanced mixers with low noise, and amplifiers hindered terahertz heterodyne low noise receiver circuit development. Available sources had extremely small output powers and required development of very low loss circuits that eliminated materials and radiation.

The Center's goal was to produce at least 1 mW at 1 THz by 1993. Oscillator power levels were in the 10s of milliwatts in 1988. Near 100 GHz, low noise GUNN diode output power was 40 mW (100 GHz) compared to slightly higher noise Si IMPATT diodes at about 10 mW (120 GHz). By 1998, TUNNETT diodes produced more (100 mW±5 mW @ 100–107 GHz) while GUNN diodes improved by 5x (200 mW at 103 GHz). Power levels improved but declined as frequency increased (i.e., 80 mW @152 GHz, 9 mW @ 209 GHz, 4 mW @ 235 GHz) [2]. GUNN diodes eventually produced 0.1–1 mW at 400 GHz in 2004. Low cost imaging systems were being developed with 200 GHz sources for security applications [3]. Today, purchased sources operate up to 1 W at 100 GHz and 10 mW at 300 GHz (http://terasense.com/products/terahertz-sources/?gclid=Cj0KCQjw4s7qBRCzARIsAImcAxacrjcx03kJlqLMfFCIqVMLAeUoDl6asbCf0PxNgdcGXXaDU8ugOXYaAnbTEALw_wcB). Terahertz can now be achieved with frequency multipliers.

Low noise receivers also required low-power passive component designs. For passives, research objectives were how to (a) decrease passive (e.g. interconnect, filters, etc.) component loss, (b) minimize electromagnetic interference (EMI), and (c) minimize and/or eliminate crosstalk effects due to close proximity within integrated circuits. My project goal in Katehi's group was to develop design methods to reduce passive circuit loss and crosstalk effects to improve the passive circuit performance.

10.3.3 Microelectronics, Microwaves and Low Power

Michigan's strong research program in microwave solid-state devices, circuits, and systems had exceptional fabrication facilities to support III-V active devices and silicon MEMS sensor research. The center's team had leading faculty and researchers on devices (e.g., P. Bhattacharya, G. Haddad, D. Pavlidis, J. East, and S. Pang), circuits (e.g., L. Katehi), antennas (e.g., G. Rebeiz), and systems (i.e., F. Ulaby) and projects for about thirty graduate students.

My research specifically focused on developing packaging solutions at the circuit level and on developing low loss miniaturized passive components such as interconnects, filters, antennas needed to build higher-level circuits. The work was extremely novel at the time with very little prior work in existence, a perfect scenario for me. My GEM MS and PhD fellowship supported the first three years to develop basic skills in *three-dimensional visualization of structures*, *test and measurement skills*, *working with different research teams* (e.g., modeling and silicon micromachining researchers from Ken Wise and Khalil Najafi's sensor group) and *learning to fabricate circuits* in a cleanroom that combined integrated circuit and silicon micromachining techniques. The challenge at the time was how to make deep wet-etched cavities using silicon micromachining techniques before deep reactive ion etching (DRIE) machines were commercially available.

Developing high frequency circuits in silicon opened a door that is commonplace today. In 1988, however, it was taboo and received much pushback from the III–V community that produced and still does the best and fastest active devices in the world.

Silicon for high-speed GHz applications had several hurdles to overcome. To conserve the very low source output power levels, losses needed to be minimized or eliminated in passives. Substrate loss contributions could be minimized by using high resistivity (>2000–3000 ohm-cm) silicon at 10s of GHz operation. Above 100 GHz, dielectric loss exceeded conductor loss as shown by its linear dependence with frequency and therefore needed to be removed. For active devices, III-V devices proved to be superior and would be co-integrated into circuits. Despite the challenges, silicon-based RF and microwave integrated circuits were born and emerged with hybrid integration using flip chip technology.

Waveguide-based circuits, formed using silicon micromachining, had very low loss [4] with no dielectric material that were very lightweight because of the use of silicon instead of metal. Extremely low loss planar interconnects were also developed called microshield with suspended metal conductors on a relatively thin dielectric membrane [5]. Low noise receivers using Schottky diodes were developed with silicon micromachined horn antennas to operate near the 100 GHz range [6, 7].

My work focused on reducing radiation and near field proximity effects due to coupling and crosstalk issues, respectively. I developed integrated packaging solutions and circuits to shield transmission lines and other passive components at the wafer-level of unpackaged circuits [8]. These integrated packages were the first of their kind and resulted in my first patent. While the high dielectric constant (~12) value of silicon was great for miniaturizing passive components (e.g., filters), it degraded printed antenna performance. Patch antennas, for example, perform best on low dielectric constant material close to air. The high dielectric constant materials inhibited the release of energy from the antenna and create undesired electromagnetic waves that travel along the surface of the substrate instead of away from it. Using Si micromachining, I synthesized the low dielectric constant environment for the antenna to achieve an optimum performance [9].

The NASA work also led to the formation of future advances in three-dimensional packaging [10], high isolation [11], and RF switch technology with incredibly low loss [12]. As a result, the area of RF MEMS emerged and gained DOD support through DARPA funding after the NASA Center ended. The switch technology, while not my research, gave exceptional performance and was a promising solution to replace transistor-based switches. There were a number of challenges that researchers were overcoming like lowering the initial high dc voltages. Ultimately, switch reliability was not competitive with other lower performing technologies.

10.4 The Information Age, Wireless Technology and High-Speed Microelectronics

In the mid-1990s, a number of technologies showed significant promise for advancement, namely fiber optic communications, wireless technology, and nanotechnology. It was timely and near the turn of the century when the Information Age was born. These were exciting times, but also turbulent times of change with many unknowns.

10.4.1 On Being a Professor

My twelve years of engineering education were influenced by scientific exploration for space and technology development for low-power mobile communications. I wanted to go into industry for a few years before becoming a professor to obtain practical experience in the field. By 1995, the needs of industry were changing dramatically. The economy was so unstable with mergers and downsizing of leading companies in my field, it seemed like a one-way street for a first-generation engineer and researcher in the field. Since I was a strong candidate for an academic job, I joined the university instead. In the beginning, I felt like I was being baptized by fire, but eventually I learned to take the heat. I never learned to like it though.

I started my first academic career at the University of Illinois at Chicago (UIC) and was then recruited to the University of Minnesota. In both cases, senior colleagues approached me and were interested in bringing RF MEMS, a hot topic at the time, to their departments applied electromagnetics programs. I was very fortunate that my research resulted in first place in the student paper competition at our flagship conference, the International Microwave Symposium. It gave me a lot of visibility.

Dr. P. Uslenghi at UIC invited me to present my work after seeing it at an antenna conference. I would eventually join his department in the Fall of 1995. Dr. A. Gopinath would also talk to me at the International Microwave Symposium about my work on passive components. It would be three years later before I would join his department in Fall 1998. My NSF CAREER award focused on combining microwave electrical signals with optical ones and he was an expert on optoelectronic devices. Prior to joining his department, he had agreed to help me on the optical portion of this work.

10.4.2 Turn of the Century Communications

US semiconductor manufacturing was strong and expensive. As companies downsized, they replaced research cleanroom facilities and associated research with more applied and development research in alignment with product lines. At the same time, commercial fabrication facilities like Taiwan Semiconductor Manufacturing Company Limited (TSMC) founded in 1987 were growing in success to perform silicon integrated circuit fabrication. As a result, fabless electronic design companies started to emerge more aggressively, leaving a few large companies in computing, like Intel, with strong fabrication capability. Communication companies, on the other hand, were still focusing on miniaturization and exploring ways to build fully integrated communication systems.

Microwave communications relied primarily on III-V materials, especially for power transistors with low noise and power amplifiers in the late 1990s. Mixed signal designs were also growing in importance. In the meantime, fabless communication companies like Qualcomm (est. 1985, San Diego, CA) and Broadcom (est. 1991, Los Angeles, CA) were advancing wireless handheld mobile phones by working to miniaturize discrete hybrid circuits and ultimately to replace them with fully integrated circuit ones. The main challenge was how to combine the different material systems (i.e., silicon, III-V, etc.) required to satisfy the different active and passive device needs. RF CMOS would start to grow more rapidly by mid-2000s after the designs were more clearly understood.

Co-integration of RF communications with computing using CMOS design is the Holy Grail. Substrate resistivity requirements conflict, however. Active circuits need low resistivity substrates (≪1 ohm-cm) and passive ones need high resistivity (>2000 ohm-cm). One of my first University Minnesota projects investigated novel methods to reduce dielectric loss. We explored the use of porous silicon to increase the resistivity in specific local region in a low conductivity substrate. Then, we developed interconnect and filter circuits to understand performance impact [13] and planar characterization circuits of the same type [14] to characterize material loss. Losses reduced considerably, but never as much as in high resistivity silicon.

Optical communications had many breakthroughs at this time and radio frequency (RF) photonics was emerging as a promising field. Integrated optical systems rapidly advanced due to the development of silicon optical microbench technology. Fiber optical systems also advanced fiber optic amplifier discoveries. The latter opened many doors for optical systems use in the infrastructure that could enhance Ethernet access in computing applications like the Internet. It provided both very low loss compared to copper electronic systems and increased bandwidth for the anticipated high number of users. A need existed, however, to interface microwave electrical signals with optical ones.

Finding ways to enable advancements in communications, optically and/or wirelessly, was a key focus for the first decade of my career (1995–2005). Computers were becoming more portable and mobile and the Internet was emerging and growing by leaps and bounds. Wireless communications technology for high volume applications was in its infancy both in understanding its potential and its designs.

Optical communications needed better electrical interfaces for microwaves signals to modulate and detect or convert optical light signals into high-speed electrical signals since computing was and is still largely electrical. This need motivated focus of my work on problems that combine microwaves with optical signal applications. It was also the focus of my National Science Foundation young investigator award, the 1998 Presidential Career Award for Scientists and Engineers.

We developed design and fabrication methods to facilitate integration of two- and three-dimensional integrated circuit systems operating at microwave and millimeter wave frequencies that provided ultra-broad band (110 GHz) operation. A visiting faculty scholar stint at LLNL resulted in a collaboration to develop low crosstalk vertical cavity surface emitting lasers (VCSELs) systems for wavelength division multiplexing for terabit storage applications Fig. 10.1 [15]. We built microwave feedline arrays into a silicon optical microbench platform that could house the laser array and associated optical fiber connections [16]. This work led to research on how to simplify high-speed integration on a single layer without wire bonds [17] for circuits requiring curved layouts with comparable performance. We also investigated how to improve the bandwidth performance of vertically integrated chips using flip chip technology [18–21] in wafer-scale packaging. Other solutions prior to ours focused on design approaches for unpackaged chips.

Reconfigurability was the next important challenge in communication system design to support rapid changes in performance requirements. We investigated the use of fluids for tuning of filters and antennas [22, 23] for radar applications where tuning and cooling were needed. This work resulted from an interdisciplinary collaboration to understand how bacteria conducts electricity, using nanowires or plasma, with a biochemist colleague [24]. Our IC fabrication capability attracted them to us and the potential for biodegradable electronics attracted my interest. Extensive time went into investigating how to grow bacterial thin films with compatible integrated circuit fabrication processes. While new biology understanding resulted, microwave test results were limited due to the inability to measure “wet” circuits in my lab at the time. This inspired my interested to measure circuit with liquid solutions and explore more bio-related applications. More later.

Fig. 10.1 On-wafer testing of 50 GHz integrated interconnects in silicon optical microbench technology for optoelectronic applications. Persons in photo: Riki Banerjee (sitting) and Rhonda Franklin (aka Drayton) (standing)

10.4.3 Shifting Gears: Microwaves Meets Nanomedicine

Nanotechnology popularity in materials emerged at the start of my career and was not understood or developed well enough to use directly in my work. I was always interested in working in this area, but it was unclear how to do so. After years of talking to a colleague who had expressed interest in working together for almost a decade, we developed an idea over lunch. Within a week, we had written a proposal that was successfully funded and we have been having fun ever since. This new topic has allowed me to blend two areas I find fascinating—materials and medicine—with my expertise in advancing communications technology.

My group and Dr. Bethanie Stadler's group are working on two problems with microwaves and nanotechnology. First is how to detect magnetic nanowires by their ferromagnetic resonance microwave properties for nanomedicine applications. Second is how to integrate copper nanowires into via technology for sub-millimeter wave applications. The second project is also exciting and includes a collaboration with another colleague, Dr. Rashaunda Henderson, currently at UT Dallas.

10.4.3.1 Magnetic Nanowires in Nanomedicine

We have been developing methods to characterize the microwave properties of Stadler's magnetic nanowire structures for nanomedicine applications. We are specifically interested in designing biolabels for cancer cell research. These labels can have different addresses (i.e., signatures) based on the type of material or combinations of magnetic materials. Labels with different addresses based on their microwave properties (i.e., ferromagnetic resonance) can be exposed to cell systems at the same time. Those cells with the labels in them can be separated and read simultaneously to determine what type of cancer feature is present. My team is creating the diagnostic circuits and algorithms that can then be used to identify the specific magnetic nanowires present. This approach will create a parallelized identification system compared to many of the serial ones currently used today.

Ferromagnetic resonance (FMR) is a natural resonance that exists in magnetic materials when a specific combination of a microwave signal and static magnetic field results in microwave absorption. We have built circuits to test chip scale versions of magnetic nanowires and have determined how multiple samples can be detected simultaneously on printed coplanar waveguide circuits in different test configurations [25]. This feasibility study confirms that multiple types of wires can be simultaneously detected. Stadler's group fabricates chips with magnetic nanowires in aluminum oxide host wafers with empty nanopores. When these nanowires are released from chip form they are used to understand nanowire behavior in the cancer cell studies [26]. We are now interested in studying how they behave electrically when released from chipform and in developing methods to introduce alignment fields to evaluate their response when they are aligned vertically but not ordered.

10.4.4 Moving Forward: Retro and New Beginnings

A new collaboration with Dr. B. Stadler and Dr. R. Henderson studies how copper nanowires, when used as vertical connections, perform from microwave (30 GHz) to sub-millimeter wave frequencies (300 GHz). This work is relatively new and promising for developing interfaces between nanoelectronics and larger scale electronics. We are interested in understanding how copper nanowires in different shapes and sizes can be developed to offer lower conductor loss to support signal power levels up to mW at sub-millimeter wave frequencies.

I am also interested in nanowarming in nanomedicine related to organ-preservation applications. Organ preservation is in high demand, low supply, and needs solutions to improve storage and use of the available supply. Exciting interdisciplinary work exists to help facilitate moving this field forward that can benefit from methods used to advance materials characterization in microwave microelectronics.

My research over the years on interconnects, integration, integrated packaging and passives—circuits and antennas, as well as material characterization has provided the flexibility to contribute to a number of application spaces. As a result, my team and I have worked on a plethora of topics that have moved the field of advanced high-speed communications forward and are blazing new trails in nanomedicine by making key connection between nanotechnology and microwave microelectronics research.

References

1. Ulaby FT (1990) Overview of the University of Michigan space terahertz program. In: Proceedings of the first international symposium on space terahertz technology, Ann Arbor, MI, pp 5–32
2. Eisele H (1998) GaAs TUNNETT diodes and InP GUNN devices for efficient second harmonic power generation above 200 GHz. In: Proceedings of the ninth international symposium on space terahertz technology, Pasadena, CA, pp 587–596
3. Gallerano BP, Biedron S (2004) Overview of terahertz radiation sources. In: Proceedings of the 2004 FEL conference, pp 216–221
4. Siegel PH (2002) Terahertz technology. IEEE Trans Microw Theory Tech 50(3):910–928
5. Weller TM, Rebeiz GM, Katehi LP (1993) Experimental results on microshield transmission line circuits. In: 1993 IEEE MTT-S international microwave symposium digest, Atlanta, GA, USA, vol 2, pp 827–830
6. Ling C, Rebeiz G (1993) A 94 GHz integrated monopulse receiver. In: Proceedings of IEEE antennas and propagation, pp 608–611
7. Gearhart SS, Rebeiz GM (1994) A monolithic 250 GHz Schottky-diode receiver. Proc Trans Microw Theory Tech 42(12):2504–2511
8. Drayton RF, Weller TM, Katehi LPB (1995) Development and characterization of miniaturized circuits for high frequency applications using micromachining techniques. Int J Microcirc Electron Packag 18(13):217–224. Third Quarter 1995
9. Papapolymerou I, Drayton RF, Katehi LPB (1998) Micromachined patch antennas. IEEE Trans Antennas Propagat 46(2):275–283
10. Henderson RM, Herrick KJ, Weller TW, Robertson SV, Kihm RT, Katehi LPB (2000) Three-dimensional high-frequency distribution networks – part II: packaging and integration. IEEE Trans Microw Theory Tech 48(10):1643–1651
11. Drayton RF, Henderson RM, Katehi LPB (1998) Monolithic packaging concepts for high isolation in circuits and antennas. IEEE Trans MicrowTheory Tech 46(7):900–906
12. Katehi LPB, Pacheco SP, Peroulis D (2000) RF MEMS for wireless communications systems. In: Proceedings of the international conference on microtechnologies, Hannover, Germany, vol 2, pp 25–27 (Invited Paper)
13. Itotia IK, Drayton RF (2005) Aperture coupled patch antenna chip performance on lossy silicon substrates. In: 2005 IEEE antennas and propagation society international digest, vol 1B, pp 377–380, 3–8 July 2005
14. Peterson RL, Drayton RF (2002) A CPW T-resonator technique for electrical characterization of microwave substrates. IEEE Microw Wirel Compon Lett 12(3):90–92
15. Rajesh RP, Steven WB, Michael DP, Michael CL, Henry EG, Rhonda FD, Holly EP, Denise MK, Robert JD, Mark EL (2003) Multi-wavelength parallel optical interconnects for massively parallel processing. IEEE J Select Top Quant Elec 9(2):657–666
16. Banerjee SR, Drayton RF (2006) 50 GHz integrated interconnects in silicon optical microbench technology. IEEE Trans Adv Packag 29(2):307–313

17. Kim H, Franklin-Drayton R (2009) Wire-bond free technique for right-angle coplanar waveguide (CPW) bend structures. IEEE Trans Microw Theory Tech 57(2):442–448
18. Franklin R, Cho YS, Kim H (2010) Ultra broadband high-speed integration techniques for circuits and systems. In: 11th IEEE wireless and microwave technology conference (2010 WAMICON), Orlando, FL, pp 1–4, April 2010 (invited paper)
19. Cho YS, Franklin RR (2010) Novel broadband through silicon via interconnect for three dimensional CPW transition. In: 2010 European microwave conference, Paris, France, pp 113–116, September 2010
20. Cho Y-S, Franklin-Drayton R (2009) Characterization and lumped circuit model of ultra-wideband flip-chip transitions (DC - 110 GHz) for wafer-scale packaging. Microw Opt Technol Lett 51(5):1281–1285
21. Cho Y-S, Franklin-Drayton R (2009) Development of ultra-broadband (DC-50GHz) wafer-scale packaging method for low profile bump flip-chip technology. IEEE Trans Adv Packag 32(4):788–796
22. Miranda G, Franklin R (2014) Fluidic tunable microstrip bandstop filter. In: 15th annual IEEE wireless and microwave technology conference (WAMICON 2014), pp 1–3, June 6, 2014, Poster Presentation
23. Murray C, Franklin RR (2013) Frequency tunable fluidic annular slot antenna. In: 2013 Antennas and propagation society international symposium (APS/URSI), pp 386–387, https://doi.org/10.1109/APS.2013.6710854
24. Liu H, Kim R, Franklin DB (2011) Linking spectral and electrochemical analysis to monitor c-type cytochrome redox status in living geobacter sulfurrenducens biofilms. CHEMPHYSCHEM, 12(12):2235–2241. https://doi.org/10.1002/chphc.201100246
25. Zhou W, Um J, Zhang Y, Nelson A, Nemati Z, Moidiano J, Stadler B, Franklin R (2019) Development of a biolabeling system using ferromagnetic nanowires. IEEE J Electromagn RF Microw Med Biol 3(2):134–142. https://doi.org/10.1109/JERM.2018.2889049
26. Nemati Z, Gage T, Zamani K, Mohammad R, Shore D, Um J, Subramanian S, Franklin R, Modiano J, Stadler B (2018) Magnetic isolation of exosomes using Fe/Au nanowires: towards an improved early detection of cancer. In: 2018 MMM-interrmag conference

Chapter 11
A Creative Path Towards Becoming Female Engineer Enabling the Next Opportunities in Computing

Jennifer Hasler

> It is brave that you *choose* to become a *female engineer*
> —Sunny Bains, 2011

Not the kind of email response one gets every day. I had known Sunny for several years, and we had worked on a couple engineering communications over the previous decade or so. The statement struck me in so many ways. Why would anyone want to be female and in engineering given how hard the road. One might try to become an engineer if you were female, but to become female if one was already trained in engineering seemed between foolish and terrifying. And yet, it is my journey.

My journey is different from the other amazing women in this book, several whom I know personally and can attest to their struggles and successes. I do not have a story of barriers overcome preventing me to study or practice engineering because of my gender. For me, if I just lived according to social norms and behaviors, I faced no apparent roadblocks other than my own abilities to master the material. And yet, that was not enough.

I was not the first engineer to transition female socially, as brave women like Lynn Conway did decades earlier, losing her career and reputation at IBM as well as her children. Other engineers who have taken a similar journey might not believe they have the right to stand with other women engineers, or speak for the journey of other women engineers. While I respect their viewpoints, my journey, as well as the journey of my spouse, daughters, colleagues, and students compels me to stand and take my place with other women engineers, weaving my own story into the larger fabric.

My path has been a creative one, walking through my journey through engineering education, my journey to becoming an engineer working on the cutting edge of computation and neuromorphic techniques, and then becoming a female engineer. Through this chapter, I wish to share with you one perspective walking through my

J. Hasler (✉)
Georgia Institute of Technology, Atlanta, GA, USA
e-mail: jennifer.hasler@ece.gatech.edu

A. C. Parker, L. Lunardi (eds.), *Women in Microelectronics*, Women in Engineering and Science, https://doi.org/10.1007/978-3-030-46377-9_11

journey. Often through this process, I remember the Robert Frost poem (The Road not Taken) with the lines

> Two roads diverged in a wood, and I—
> I took the one less traveled by,
> And that has made all the difference

11.1 A Creative Path Through Traditional Engineering Education

Among my family, I was the first practicing engineer. I remember considering engineering after two UCF engineering students came into our middle-school gifted classroom to give an overview about engineering. As I already enjoyed math, science, and computer programming, the direction was fascinating. Only decades later did I realize the personal importance that both engineering students were female. At a young age, I knew I was different, and not only did I connect well to girls rather than boys, but I both wanted to be them and viewed myself through these lenses. I was the eighth grader in chorus who still sang first soprano even after my voice broke; my mind knew where I was supposed to sing.

I always enjoyed Math and Science, and yet, I really enjoyed creating new things. Math and Science areas just seemed like a natural place to be creative. I know having a creative child sometimes created stress for my parents, as a creative child might not have efficiently followed directions. I was always curious what else could be done. I took to having Legos from the time I was 5 years old, and would spend hours creating new things everywhere. Whether it was developing new Lego spaces, or creating music on the keyboard, creating a unique theological perspective, seeing history in a different way, or creating a video game, I enjoyed the creative process. I personally enjoy the creative aspect of cooking and attempting to create something better every time I tried. Engineering just seemed natural as a creative opportunity, an opportunity to paint with the canvas of transistors an array of supporting opportunities. Students who have studied with me likely see these aspects, whether or not they themselves personally relate. I have also always enjoyed teaching anyone who would listen, as my daughters can attest, often with eyes rolled.

My family was always supportive of my directions, although they did not know how to help me throughout my journey. My immediate family, as well as my extended family, was very close growing up, and always with high expectations. My family came closest to engineering with my great-grandfather starting an electrical contracting business in the early 1900s, a business that lasted until my grandfather and uncle sold it in the late 1950s. My great-grandfather had my grandfather start in the new area of electrical engineering in Cornell, but that only lasted a semester or two before eventually he took his place in the family business. My grandfather saw my graduation B.S.E. and M.S. from Arizona State University (ASU) in August 1991, two months before he passed away.

My passion for doing engineering pushed me to take a unique path through my education. My younger brother, who earned his B.S.E. in Biomedical Engineering at USC on his way to earning his medical degree (at USC) speaks of it as "manipulating the (educational) system." I maximized my high-school opportunities in math and science, and I was looking to do so more. Both of the major universities in Arizona (Tempe and Tucson) had a program for high-school students to live on campus and take a college course during their five-week summer semester. One school locked down any opportunity to real engineering courses until all general education requirements were completed. The other school, while confused initially, was at least willing to discuss the opportunity. I remember the program coordinator saying she would give someone a call, and a meeting was set up with Dr. Kelly, an associate dean of the engineering college at ASU. I had the opportunity to take linear circuits (ECE 301), a junior-level engineering course, between my junior and senior high-school years. I took to the subject immediately. I had a colleague in the class over lunch *accuse* me of really enjoying the compressed 5-week course, as opposed to others just trying to survive. Although I denied it, she was absolutely correct. After taking this course, I had the opportunity to take other electrical engineering courses in fall semester, and I would continue in spring semester.

The opportunity that came in spring semester (1987) would be the first key pillar that would shape my research career. Part of my interest in electrical engineering was the opportunity to build better computer games and graphics. I studied designs and schematics for hardware, and learned assembly language by ninth grade. If one could build more memory or add another processor, one could make a far better game. So what could be better than making new computer chips that one could make far better games?

I noticed that a graduate level course in digital VLSI design (ECE 525) was being offered that spring semester at a time I could manage with my high-school schedule. I had a lot of the background by that point, so inquired with the professor, Dr. Lex Akers, if taking such a course would be possible. Even given the uniqueness of the request, eventually it became possible, although through a blended course that included the VLSI design course and the senior-level digital design course (ECE 425) that I independently learned. I immediately took to the material, although I realized there were times I was missing some of the prerequisite language and material. I just pushed harder given the challenge. I finished my first graduate course a couple weeks before I graduated high-school.

The background in digital VLSI design prepared me for starting my research trajectory. One day on campus in late April as the digital VLSI course was nearly finished, a few Ph.D. and M.S. level graduate students and I were in Lex Akers' office. The discussions moved towards building custom circuits for the new hot field of Neural Networks (NN). I did not get everything in that first discussion, and yet, I could see there were opportunities for a range of architectures. I remember Lex Akers saying, "there will be a Nobel prize in this area someday." Figuring out how to make physical systems compute and learn like the human brain seemed like goal worth pursuing. *It still is a goal worth pursuing*.

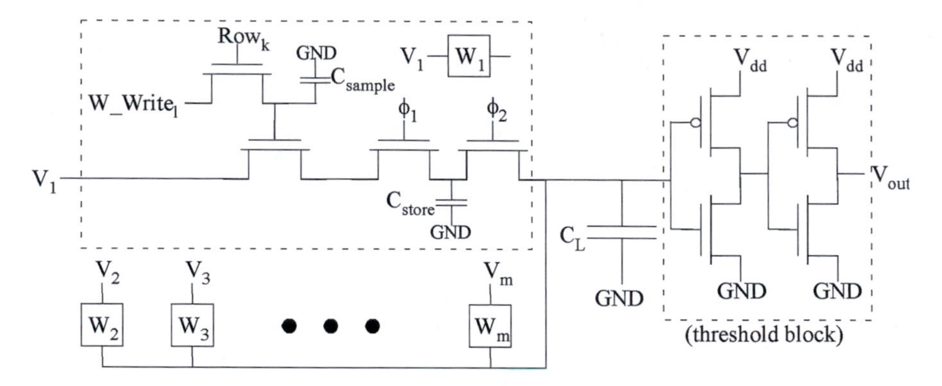

Fig. 11.1 An initial charge-summing neuron and synapse implementation for a Neural Network (NN). The neuron had m inputs ($V_1 \ldots V_m$), and m dynamically stored weights ($W_1 \ldots W_m$) that were dynamically stored voltages on C_{store}. The charge was sampled into initial capacitors that would aggregate the resulting charge on a single node (C_L)

My first publishable results emerged that fall semester. Lex Akers made a challenge to try to figure out how to make as simple a circuit that could compute a NN. Realizing that the inputs and outputs of a NN typically are binary, where whole classes of NN at the time had binary input and output signals, a charge-summing based structure could implement this design (Fig. 11.1) with a few transistors when dynamically holding weighting voltages [1, 2]. A couple of years later, I would learn that Yannis Tsividis published the first version of a charge-summing structure [3], but with some considerable differences. Both approaches had their region of application. Individuals are still replicating these directions today (e.g. [4]).

A colleague of mine, Mark Walker, a Ph.D. student, mentioned he really liked the design, and then gently asked me "I don't see anything digital in this design. It looks analog to me." Although I know he was being encouraging, it felt insulting. It was not an analog design. I don't do *that* stuff. I eventually understood that analog design at its essence was efficiently using transistor devices for computation. And yes, I have taught analog design at Georgia Tech for the last two decades, and yes, I approach analog design differently.

11.2 A Creative Path Towards Neuromorphic Engineering

Being at Caltech was magical, particularly getting to work with, and be mentored by, Carver Mead, for five years. Caltech was a place that marked its reputation on winning Nobel prizes rather than on football accomplishments (they stopped fielding a team in 1993) or basketball accomplishments [5]. There would be spontaneous celebration when a Nobel Prize winner from the campus was announced, such as Rudy Marcus in 1992, whose lectures on statistical mechanics I attended in

1994–1995. There would be frequent sightings of famous scientists, like Steven Hawkings, searching for books in the library. The informal meetings, several meetings over food, and discussions in the mountains north of Pasadena would have tales of individuals talking about the legends of science, individuals who were their personal friends. I made sure never to lose the sense of amazement and wonder of this special place, and yet, from the first day I stepped on the campus, I reminded myself never to act as if you do not belong.

I started my Ph.D. degree at Caltech in Computation and Neural Systems (CNS) in the fall of 1992. During my time at ASU, I knew I had a strong passion for teaching and research and felt called to make a difference in the lives of graduate students working through the process. By the time I finished my joint B.S.E. and M.S. degree at ASU, I had been involved in multiple IC designs primarily focused at many different aspects of NN architectures (Fig. 11.2), circuit implementations, and analog memories (e.g. [6, 7]). It took two application attempts to finally be accepted at Caltech, and I grew considerably because of the experience. And although I graduated from a much lower ranked school than my other Caltech colleagues, I never let those issues get in the way of contributing as an equal while at Caltech.

During one of his group meetings in March 1992, I met Carver Mead for the first time and immediately sensed his positive energy. I knew about his legendary career, as well as knowing how he created such a positive environment for everyone who worked with him. After having experienced a typical graduate school process, I was already passionate with a calling so that I could become faculty to make the graduate student perspective more positive. It was Carver's reputation that drew me to be part of his group. Until we had met, I was not certain if our technical areas would fully line up, but I was already willing to move my technical goals around. In many ways, things worked out better than I imagined. I am thankful for the way that Tobi

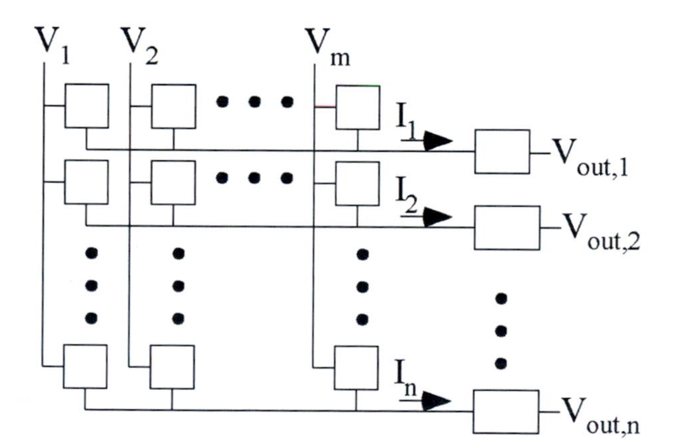

Fig. 11.2 Physical NN implementations tend to be optimally implemented as a mesh architecture where synapse elements are the center of the mesh. The inputs ($V_1 \ldots V_m$) are locally input through global broadcast along columns, and the outputs ($I_1 \ldots I_m$) are summed along the row lines and go through neuron elements for the resulting output voltages ($V_{out,,1} \ldots V_{out,n}$). This structure effectively creates a crossbar architecture

Delbruck, who defended his Ph.D. thesis just after I started in 1992, was a supportive mentor for me. Our group was an amazing group of people (Brad Minch, Chris Diorio, Kwabena "Buster" Boahen, Lena Peterson, Shih-Chii Liu, and Rahul Sarpeskhar) who mostly remained together throughout my entire time as a graduate student. I was always the most junior and youngest student in the group, even on the day I defended my dissertation in 1997.

Neuromorphic Engineering is a unique term, starting by the joint belief of John Hopfield, Richard Feynmann, and Carver Mead, that one expected overlap between electrical engineering, computing, and neuroscience but having little idea how to build something in this area. In classic Caltech tradition, they started a class so they would understand the field. This year-long course in 1981 became several courses and eventually became a degree program, Computation and Neural Systems (CNS). By the time I started in 1992, this Caltech community had a decade of experience. I understood the wider neuromorphic engineering community before I arrived at Caltech, having published and being at conferences for years, understanding the perspective within the family, as well as the outside perspective.

Neuromorphic Engineering looks to the computing opportunities in neural systems, implement those computations in synthetic systems, and in the process of building these computations, one asks new neuroscience questions. As Carver Mead said,

> "if you build it, you understand it. And if you understand it, you can build it"—Carver Mead

One can see Richard Feynman's influence. The computational advantages came from the great potential by physically based computation over digital computation:

> "taking all the beautiful physics that is built into…transistors, mashing it down to a 1 or 0, and then painfully building it back up with AND and OR gates to reinvent the multiply." —Carver Mead [8]

One major reason for studying the computation of the human brain (or other nervous systems) is the impressive amount of computation performed with minimal power. The human brain requires 20 W average power, roughly 20% of the resting human body's energy consumption. Carver would hypothesize that physical (e.g., analog) computation would be at least ×1000 lower energy compared to digital computation based on transistor counting arguments [8]. I would later be part of experimentally proving this hypothesis [9], and repeating that proof many times later (e.g. [10]).

When starting in Carver's lab, it was clear that that lack of a long-term memory device limited everything in the research community. By the early 1990s, everyone understood that the lack of an analog memory, particularly for storing network weights, was the primarily limitation for any analog implementation to move forward. The lack of a memory was *the struggle* that defined the field. Having both worked in this field for five years and published in these areas (e.g. [6, 7]), I was painfully aware of the need. The lack of an analog memory threatened to make this field practically irrelevant.

Carver saw the opportunity of a Floating-Gate (FG) device potentially being that solution, and yet, when I had started at Caltech, both academic and commercial attempts to make analog memory and computing devices in Si remained challenging. Three of us, Brad Minch, Chris Diorio, and myself all looked at different and related aspects of these questions. While many stories could be told of these days, all three of us benefited greatly from this collaboration decades later into our career. The presentation of The Single-Transistor Learning Synapse (Fig. 11.3) that I gave at EMBS (October) and NIPS (December) in 1994 [11], and at ISCAS in 1995 [12], demonstrated the first analog CMOS computation, a long-term memory element that could be embedded into the computation and adaptation. One could both adapt a crossbar of these elements, as well as measure or program a single device without affecting any of its neighbors. These devices enabled a wide range of creative translinear circuits [13], as well as adaptive computing elements [14].

In 1997 I re-emerged into the traditional engineering areas starting as an assistant professor at Georgia Institute of Technology (GT) after being immersed in the amazing Caltech culture for five years. The potential of FG techniques for circuits and systems looked transformational. And yet, as assistant professors sometimes find out, most were not so excited with these "new" concepts. May times in my first years at GT, I was told not to teach or do research in that "subthreshold stuff." Most did not even want to hear anything about FG or neuromorphic concepts, techniques that they thought would likely never become commercially relevant. A few things changed over two decades.

In spite of these issues, I started building my research group, developing my teaching pedagogy, building bridges with other faculty, and staying centered to the opportunities I knew would materialize. After working with Carver for five years coupled with my ASU experiences, I knew how I wanted to build my research group

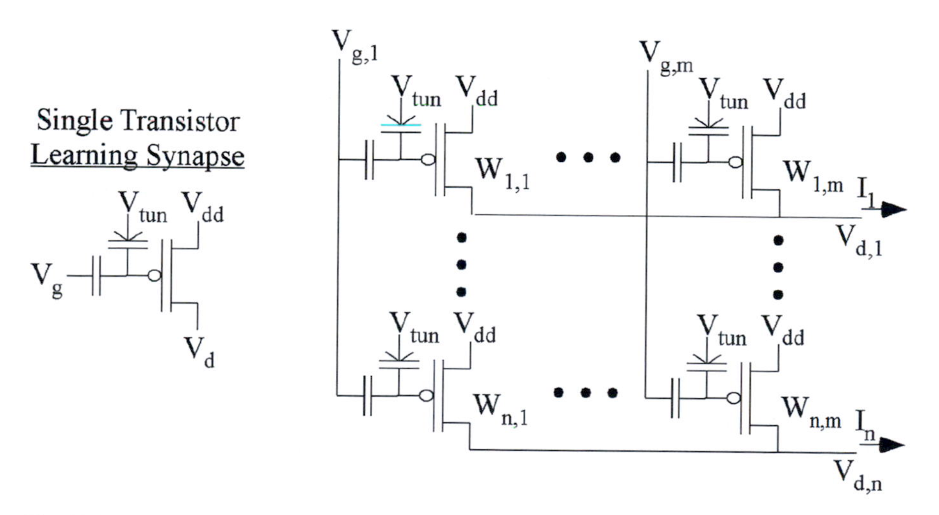

Fig. 11.3 Single Floating-Gate (FG) element used as a single-transistor learning synapse element. This structure satisfies the architecture in Fig. 11.2, enabling long-term storage, computation, and adaptive capabilities in a single-transistor element

and its resulting culture. The goal was to empower each individual as well as our collective community. In my fifth year at GT, I remember looking around at my group meeting being amazed at this community of more than 20 graduate students, being thankful at how God brought everyone together, and somewhat wondering what would I do now with this community.

I had the opportunity to work with and co-teach with Phil Allen, one of the legends of traditional IC design, where we both developed a respect for each other's perspective as well as became colleagues. This collaboration gave me the perspective and language for designing FG techniques towards classical analog circuit approaches (e.g. [15, 16]). Mismatch limits the performance of both classical commercial digital and analog designs. Having a programmable FG element in standard CMOS provided a method to minimize (or eliminate) circuit mismatches.

In the neuromorphic community, the question everyone feared and expected throughout the 1980s and 1990s was "Why not implement your neural algorithm using DSP chips?" Even today similar forms of this question are feared by many people. I developed bridges with the DSP community, working with many of the founders in DSP, Ron Schafer, Jim McClellan, and Russ Mersereau, in a group that was the best DSP group anywhere in the world while I was an assistant professor. One of my proudest moments before earning tenure was Ron Schafer introducing me to his colleague and stating I belonged to their group, that I was one of them. The feared question no longer was ever asked as we directly addressed these questions as part of our research. I started mentoring other faculty starting my third year as a professor, an action that drew significant criticism by other faculty, and yet created additional collaborations and momentum. These efforts started an over 15 year research effort in large-scale Field Programmable Analog Arrays (FPAA) enabled through the programmable and configurable opportunities of FG devices [17]. The FPAA devices opened opportunities to understand the nature of analog computing [18], including foundational work on analog numerics [19], abstraction [20], and architectures [21].

Neuromorphic engineering did not appear as a primary focus of my research, as it had nearly zero interest in the academic and funding communities for my first decade at GT, and yet this research continued throughout this timeframe. The development of channel transistor models of biological channels and computational models of neural dendritic trees were two key neuromorphic developments that arose during these days. The result of this work showed a path towards building cortex while showing additional 100,000x energy efficient improvements over analog computing techniques [10, 22]. A lower-bound estimate of the computation of the human brain would be the equivalent of 10,000 IBM Sequoia supercomputers. Neural computation must be energy constrained, as any inefficiencies allowed requires a higher calorie intake.

Both opportunities started with my collaboration with Steve Baer (Math department at ASU), whom I took my first computational neuroscience class. When taking his course (Fall 1991), we started connecting transistors and biological channels, both in physics and in nonlinear dynamics. I would continue to work through this question over the following decade. Effectively, if we could have worked with

Hodgkin and Huxley when they were doing their Nobel Prize winning work in the 1950s with current knowledge of transistor physics, what circuit model would we have developed? A decade later, we presented this transistor model on a Saturday 9 am morning meeting at the Telluride Neuromorphic Engineering Workshop in 2002 (followed by later papers [23, 24]), after Ethan Farquhar and I worked day and night on getting all of the nonlinear dynamics to match the original Nobel Prize winning data (Fig. 11.4). This model was the first serious rethinking of the Nobel Prize winning model, and has been predictive of biological physics when applied to synapses [25], dendrites [26], and networks [27].

John Lazzaro, an alumni of Carver Mead's group at Caltech, presented initial thoughts of building HMM classifiers at NIPS 1996 [28], effectively demystifying heavily guarded techniques used for speech recognition. And yet, listening to John's presentation, many opportunities for performing these computations were completely unaddressed. A few years later, I had discussions on the same day about Si implementations of dendrites based on the channel-model efforts, as well as discussions on an all-analog architecture for phoneme detection, which included Hidden Markov Model (HMM) classification. There was a moment realizing that the very circuits for variable-diameter active dendrites were very similar to the potential circuits for implementing HMM classifiers without some of Lazzaro's earlier circuit concerns [28]. The work was initially published in 2004 [29], and Lazzaro's proposed yes/no word spotting problem was solved using biologically modeled neurons utilizing dendritic coincidence detection [30]. Incorporating dendritic computation significantly increases the energy efficiency estimates of neuromorphic computing (×100,000), opening opportunities for even more energy efficient Si computing (e.g. [10]).

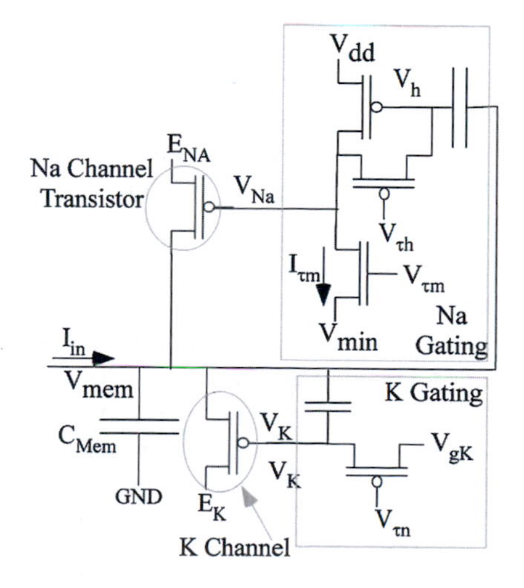

Fig. 11.4 Classical Transistor Channel implementation of Hodgkin and Huxley's original neuron measurements

11.3 A Creative Path Towards Becoming a Female Engineer

Between multiple faculty interviews at Caltech in the spring of 1997, a few unusual messages started appearing around the Internet about Lynn Conway's background. From my digital VLSI background and working with Carver Mead for nearly five years, I had a huge respect for her early work co-founding the field of digital VLSI design with Carver [31]. Investigations showed her contributions at IBM in the 1960s, including as an inventor of out-of-order instruction set execution. The reason it was hidden also became known: Lynn was fired from IBM in 1970 because a surgical gender change. I could do nothing else that day other than read everything available on this topic. Many cryptic comments over the previous four years made about Lynn started to be clear. Lynn believed no one knew, but apparently two decades later, many people eventually knew and yet had no issues with her history. The topic was personal for me, and more than the obvious technical connection and respect. As I had a faculty position starting soon, I tried to not think too hard about this news.

Ten years later, six of my Ph.D. students and I went to the annual Circuits and Systems conference in New Orleans. A couple of weeks earlier, I had obtained the book *She's not there* [32], and I could not put it down. In seeing the many parallels of a faculty member transitioning while preserving their family, I saw for the first time in my life, that my story might actually matter to someone else. And yet, how could I possibly disrupt everyone and everything that has been built? How could I survive if I did not do something? If I did something, what would be left?

The process for an assistant professor getting tenure is difficult. Years after faculty get tenure, they believe it was not that difficult, much like a few years after graduation all of the struggles of being a student have faded away. After submitting all of my final paperwork for tenure in Sept 2002, I could finally let my soul fly again. And she flew and cried out for authenticity. My faith moved from walking through the tenure building process to seeing the wider world. I was still as busy as before I got tenure in Spring 2003, now mentoring a group of over twenty graduate students while incubating a startup company, GTronix, that was co-founded with three of my first four Ph.D. students. And yet the heaviness of the process decreased.

The startup company did quite well, so well that it was the first company funded by the top-tier venture capitalists along Sand Hill Road in Silicon Valley. Being the first company, most of the company was moved to the Fremont, CA area of Silicon Valley during the summer of 2004. The Fremont to San Jose area was the center of the transgender community in the bay area, and Fremont, CA was effectively the center of a famous murder and trial of Gwen Araujo [33]. I would be traveling to Silicon Valley every month until GTronix would be eventually acquired by Texas Instruments in the summer of 2010.

As my soul became alive and my faith grew, I could feel a call towards authenticity. Many believed transition and faith were separate. In my case it was precisely because I felt God was calling me out towards authenticity that I needed to transition [34]. I know now that without transition, I would have experienced catastrophic health effects due to the stress of suppressing myself.

A key transition aspect was making sure my family would be intact through the process. Although such transitions with intact families are becoming more common today, a decade ago the number of examples were few, particularly when children were involved. My timing was not entirely my choosing, because in August 2010 my Department Chair (fortunately not at GT any longer) outed my transition to my entire department. I still postponed this process publically because I knew my family required more time to handle this transition. I officially transitioned in January 2012 at GT, even though I had practically transitioned significantly earlier. Today people often ask me how my family is doing. My response sometimes is "normal, and I am thankful for normal." I have personally walked with many individuals who transitioned and lost so much, and as a result, I am thankful for normal.

As part of my transition, I opened myself up to a few things I enjoyed doing. A few months after my official transition at GT (Jan 2012), a teaching and education group offered a book reading club (with afternoon snacks) on an educational topic. I would get an opportunity to know a different part of the GT campus, as this community was majority female in a technical school, and I was only one of a couple of engineers to ever participate. This first semester we would read a book, *My Freshman Year*, relating the story of an anthropology professor at Northern Arizona University (NAU) using her sabbatical year to immerse herself in the foreign culture of first-year undergraduate students [35]. She noticed that faculty and students walked and lived in different spaces, and would rarely interact. Most faculty have forgotten what it was like being a student, and more likely are somewhat terrified of their students. I could immediately see the appeal of such an adventure, and six years later I would do something similar for my first sabbatical semester (Fall 2018) in 21 years.

Fall 2014 I started a new adventure taking classes part-time at Emory, in a process that would lead to my pursuing a Masters of Divinity (MDiv). This journey was partially a journey to reconnect to a student perspective, as well as honor a promise I made to myself during my ASU days that I would take such classes. I graduated with my MDiv degree in May 2020. Along the way, I became very involved in campus ministry, including being the Faculty advisor for the Methodist community. My original trajectory to be faculty was a form of a calling. This journey has developed my call towards caring for the entire campus community, growing out of the original call towards graduate student mentoring. A call changes as we individually grow in authenticity. I am fortunate to have had multiple strong female role models, including a female pastor (Barbara Riddle) as part of our new church founding (Tuskawilla UMC founded in 1980). The multiple aspects of this seminary experience would be a book in itself.

These directions directly impact my current research directions. These directions have re-energized a passion for undergraduate teaching as well as ethical graduate student mentoring. And yet, with the potential successes in neuromorphic engineering and understanding the human brain, the resulting interdisciplinary questions bring in understanding from philosophy. The new opportunities in physical computing [18], which includes analog, neuromorphic, and quantum computing, bring new data and perspectives to address questions that philosophers and theologians raised for centuries and that they have only dreamed they could understand. When we build artificial or robotic systems, we effectively hard-code their purpose and function. And yet, if we build human cortex, embody it, and train it over years of data, will it act human, particularly as experienced by other humans? Will there be a soul to the machine? Over the next few decades, technical communities will directly have to address these questions, and I look forward to these opportunities. As we continue to augment our reality, as we already have smart phones attached to us, who do we become, and how do we connect to each other's humanity in these new spaces. Those who work with youth already struggle with these realities [36].

11.4 Concluding Thoughts

April 30th, 2019, I woke up recovering from gender confirmation surgery. I realized I crossed my personal last threshold in this journey towards being a female engineer. I cannot help but remember Sunny's email, and my struggle that some might believe moves me towards a lesser status. And yet the journey has not concluded, where there is so much of the journey in front of me. One thing I learned from my mother, who passed away in October 2019 after a heroic fight against Parkinson's, is to continue moving towards all of these opportunities. Nothing seems to more honor her life than to live towards these new opportunities. I look forward to the next thirty years of my career, and look in amazement on technology's positive impact on all of humanity.

We live in amazing times as seen both in technology, as well as our understanding of the wide diversity of human expressions, where each accelerates each other. As the pace of innovation continues to increase, humanity already becomes challenged with a different mode of being, a mode of being where one expects huge changes in one's lifetime. In my grandparents lifetime there was significant change, but not so fast it could not be absorbed. My daughters, both likely engineers, will likely live in a world where we always expect significant improvements in technology within their generation. Every human endeavor will change as a result of technology's change, including areas such as religion, where the human struggle to reach for outer meaning in the world, will adapt with the changing technology. Things that do not change will become less relevant.

References

1. Hasler P (1988) Implementing practical neural networks in silicon. In: Wescon conference, vol SS-1, Anaheim, CA, pp 1–7
2. Walker M, Hasler P, Akers L (1989) A CMOS neural network for pattern association. IEEE Micro 9(5):68–74

3. Tsividis Y, Anastassiou D (1987) Switched-capacitor neural networks. Electron Lett 23(18):958–959
4. Bankman D, Murmann B (2016) An 8-bit, 16 input, 3.2 pJ/op switched-capacitor dot product circuit in 28-nm FDSOI CMOS. In: IEEE Asian solid-state circuits conference
5. Greenwald R (2007) Quantum Hoops. In theaters starting, Nov 2, 2007
6. Hasler P, Akers L (1991) A continuous time synapse employing a refreshable multilevel memory. In: International joint conference on neural networks, July 1991, pp 563–568
7. Hasler P, Akers L (1992) Circuit implementation of a trainable neural network using the generalized Hebbian algorithm with supervised techniques. In: International joint conference on neural networks, June 1992, pp 160–165
8. Mead C (1990) Neuromorphic electronic systems. In: Proceedings of the IEEE, no 78, pp 1629–1636
9. Chawla R, Bandyopadhyay A, Srinivasan V, Hasler P (2004) A 531 nW/MHz, 128 x 32 current-mode programmable analog vector-matrix multiplier with over two decades of linearity. In: IEEE custom integrated circuits conference, October, pp. 651–654
10. Hasler J, Marr HB (2013) Finding a roadmap to achieve large neuromorphic hardware systems. Front Neurosci 7:118, pp 1–29
11. Hasler P, Diorio C, Minch BA, Mead CA (1994) Single transistor learning synapses. In: Tesauro G, Touretzky DS, Leen TK (eds) Advances in neural information processing systems 7. MIT Press, Cambridge, MA, pp 817–824
12. Hasler P, Diorio C, Minch B, Mead C (1995) Single transistor learning synapse with long term storage. IEEE Int Symp Circuits Syst 3:1660–1663
13. Minch BA, Diorio C, Hasler P, Mead C (1996) Translinear circuits using subthreshold floating-gate MOS transistors. Analog Integr Circ Sig Process 9:167–179
14. Hasler P, Minch B, Diorio D, Mead C (1996) An autozeroing amplifier using pFET hot-electron injection. IEEE Int Symp Circuits Syst 3:325–328
15. Srinivasan V, Serrano GJ, Gray J, Hasler P (2007) A precision CMOS amplifier using floating-gate transistors for offset cancellation. IEEE J Solid State Circuits 42(2):280–291
16. Srinivasan V, Serrano G, Twigg C, Hasler P (2008) A floating-gate-based programmable CMOS reference. IEEE Trans Circuits Syst I 55(11):3448–3456
17. Hasler J (2019) Large-scale field programmable Analog arrays. In: IEEE proceedings
18. Hasler J (2016) Opportunities in physical computing driven by analog realization. In: IEEE International conference on rebooting computing, pp 1–8
19. Hasler J (2017) Starting framework for Analog numerical analysis for energy efficient computing. J Low Power Electr Appl 7(17):1–22
20. Hasler J, Kim S, Natarajan A (2018) Enabling energy-efficient physical computing through analog abstraction and IP reuse. J Low Power Electron Appl 8(4):47
21. Hasler J (2019) Analog architecture complexity theory empowering ultra-low power configurable analog and mixed mode SoC systems. J Low Power Electron Appl 9(1):4
22. Hasler J (2017) A roadmap for the artificial brain. In: IEEE spectrum, p 27
23. Farquhan E, Hasler P (2004) A bio-physically inspired silicon neuron. Int Symp Circuits Syst 1:309–312
24. Farquhar E, Hasler P (2005) A bio-physically inspired silicon neuron. IEEE Trans Circuits Syst I 52(3):477–488
25. Gordon C, Farquhar E, Hasler P (2004) A family of floating-gate adapting synapses based upon transistor channel models. Int Symp Circuits Syst 1:317–320
26. Farquhar E, Abramson D, Hasler P (2004) A reconfigurable bidirectional active 2 dimensional dendrite model. Int Symp Circuits Syst 1:313–316
27. Brink S, Nease S, Hasler P, Ramakrishnan S, Wunderlich R, Basu A, Degnan B (2013) A learning- enabled neuron array IC based upon transistor channel models of biological phenomena. IEEE Trans Biomed Circuits Syst 7(1):71–81

28. Lazzaro J, Wawrzynek J, Lippmann R (1996) A micropower analog VLSI HMM state decoder for wordspotting. In: NIPS'96: Proceedings of the 9th international conference on neural information processing systems, pp 727–733
29. Hasler P, Smith P, Farquhar E, Anderson D (2004) A neuromorphic IC connection between cortical dendritic processing and HMM classification. In: IEEE digital signal processing workshop, Aug, pp 334–337
30. George S, Hasler J, Koziol S, Nease S, Ramakrishnan S (2013) Low-power dendritic computation for wordspotting. J Low Power Elect Appl 3:78–98
31. Mead C, Conway L (1980) Introduction to VLSI system design. Addison-Wesley, Reading. ISBN 0-201-04358-0. Early drafts found at http://ai.eecs.umich.edu/people/conway/VLSI/VLSIText/VLSIText.html
32. Boylan JF (2003) She's not there: a life in two genders. Broadway Books, New York
33. One summary overview would be *A Girl Like Me: The Gwen Araujo Story*, a biography television film, premiered on Lifetime Television in 2006
34. Hasler J (2017) Must be authentic to be prophetic. Contributed book chapter to Rainbow in The Word: LGBTQ Christians Biblical Memoirs
35. Small C (Nathan R) (2005) My Freshman year: what a professor learned by becoming a student. Cornell University Press, Ithaca
36. Zirschky A (2015) Beyond the screen: youth ministry for the connected but alone generation. Abingdon Press, Nashville

Chapter 12
From Silicon to the Brain Using Microelectronics as a Bridge

Alice Cline Parker

12.1 The Foundations

12.1.1 My Early Childhood

I can still recognize the smell of an animal lab and the sounds of the centrifuges whirring away no matter where I am because those are the earliest memories of my father's lab in the Public Health Building in Birmingham, Alabama. My father had gone to Alabama to teach and do research in the Medical School at the U. of Alabama, even though he was a PhD, and not an MD, a distinction that was very clear even to me as a child. I was born there in Birmingham, and my birthplace and early childhood influenced what I have become and who I am. I was surrounded by my father's influence.

My father was perfecting a blood test for prostate cancer, and studying the role of nutrition in fighting cancer. He raised chicks in the basement to demonstrate proper nutrition and choice of feeds, and we ate fresh eggs produced by the adult chickens. He had also raised hogs to demonstrate how to provide proper nutrition to produce lean animals, but my mother named all the hogs, and they came when they were called by name, so my father sold the hogs alive and they were trucked away. This was before I was born so I only heard the stories about the hogs or I might have become a vegetarian. I did inherit my father's interest in the natural world, in animals and plants and all living things.

My earliest memory of electronics was of vacuum tubes, the type found in television sets in the 1950's. This was followed shortly by transistor radios, marketed widely by the number of transistors contained in the radio: 5 transistors, 8 transistors, and then more. Having a scientist father and an older brother exposed me to

A. C. Parker (✉)
Department of Electrical and Computer Engineering, University of Southern California, Los Angeles, CA, USA
e-mail: parker@usc.edu

A. C. Parker, L. Lunardi (eds.), *Women in Microelectronics*, Women in Engineering and Science, https://doi.org/10.1007/978-3-030-46377-9_12

Alice age 6 (Source: Alice Parker's personal files)

My Father, Joseph K. Cline, in his Lab in Birmingham (Source: Alice Parker's Personal Files)

more science and technology than I would have seen otherwise if it were up to my musician mother. There were few women braving STEM careers when I was growing up. Fortunately my first pediatrician was a woman (Dr. Ruth Berry) and I was named for an anesthesiologist who boarded with us when I was born, Dr. Alice McNeal, now listed in the Alabama Anesthesiologist's Hall of Fame. My father trained his vastly overqualified secretary Margaret Anderson Wilkinson to do lab work, and her name started showing up on papers he published. I am pretty sure she did much of the writing as well, because he avoided writing whenever possible.

My idyllic childhood crashed in flames into poverty when my mother divorced my father so that she could have the freedom to achieve her dreams of being a writer and composer. Apart from his appearing at holidays and popping up seemingly unannounced to set me up for science projects, I only saw my father when I visited him in the summers. I still persisted in my love of science, and gained some recognition and confidence with science fair awards on hydroponics and crystal growing, all with supplies and instructions from my dad.

Living in the countryside, with very little intellectual stimulation except books, my world turned inward. My mother quickly realized she could not balance her budget on the child support and the $0.60 she charged for piano lessons, and enrolled in college when I entered first grade. A few semesters later, she was required to take a psychology course. I stumbled onto her textbook, Psychology and Life, and it opened my eyes to anything involving the brain. I was fascinated by the case studies of mental illness and brain disorders. Between that text, National Geographic magazine's stories of Jane Goodall (my biggest role model) and the chimpanzees, a children's biography of Marie Curie, and the cool electronic toys my brother had been given (e.g., a crystal radio), my future path was set in motion.

12.2 Moving into Engineering

While my father influenced me greatly, high school teachers, one in particular, had a huge impact. His name was Commander Loftus (because he was retired from the U.S. Navy), and like all Navy officers, had to have a degree in engineering. He taught physics, and, in his class, I learned about electricity and magnetism, which gave me background for my future. Commander Loftus encouraged me and recommended me for a college scholarship in Electrical Engineering. I needed the money because there was nothing set aside for me since my older brother was in college and spending my mother's available financial resources. In fact, by the time I set out for college, my father had resigned from his position as a research chemist over an ethical disagreement (he was right), and could not find employment at age 59 with a PhD in Chemistry. So I applied for the engineering scholarship. The college scholarship committee interviewed me, asking some probing questions about whether I had a boyfriend, since they had perhaps not seen a woman become an engineer before, so I asked them why they had inquired. The response was that they wanted to see if I was normal. It was many years before the implications of their question made sense to me. It was Alabama in the mid 60s.

I won the scholarship and set out for North Carolina State University (near my father) determined to conquer Electrical Engineering. My sophomore year of university, I entered the huge auditorium-style classroom for my first EE lecture, and was shocked. There was one other woman, and we were alone in a sea of about 120 male faces. Fortunately, my brother was very social and I had grown up surrounded by males more or less my age always hanging around our house where there was a front yard (almost) big enough for touch football, a place to build parallel bars for gymnastics practice, and tall trees to hold a tree house just large enough for a couple of boys and a comic book collection. So I felt comfortable in the EE classroom, although there was little social interaction with the male students. I had a goal and the goal was to get an engineering degree so that I could always find employment, a goal shaped and made urgent by my experiences with poverty. Again, my encouragement came from a teacher, Wayland P. Seagraves, beloved professor who taught that first EE class.

Prof. Seagraves encouraged me to enter a research paper, "Digital Analysis of Speech," into an IEEE student paper contest, and I was able to present my findings at an IEEE student conference in Florida. Seagraves' mentorship was invaluable. I even took a human biology course (designed for nursing students since engineers did not study human biology). Although I did interview for a couple of engineering positions, including Bell Laboratories, I won an NSF Fellowship and chose to attend Stanford University as a masters student, where Prof. Michael Arbib had been teaching and researching models of the brain, according to Stanford's glossy faculty brochure. Had he been at Stanford when I arrived, instead of having moved to U. of Mass., my career trajectory might have been very different.

12.3 Introduction to Brain Models and Letting the Brass Ring Go By

At Stanford, I was introduced to brain models, including the perceptron, and the Hodgkin–Huxley equations. I wanted to explore more in this direction, and could not find anyone to study under at Stanford. My personal life took a left turn when I followed my first husband back to North Carolina, where there did not seem to be anyone to advise me at Duke, U. of North Carolina or N.C. State that was both engineering and brain focused, so I continued Electrical Engineering studies at N.C. State, and read whatever I stumbled across that was brain related. I held a summer internship at Corning Glassworks, where I perfected a tiny but very important piece of technology that could filter out red blood cells from microscopic slide images so that the machine could count white blood cells differentially, according to white cell type. An N.C. State EE professor promised me funding in the fall to continue that work, but when fall arrived, there was no funding for me to continue anything biomedical as a dissertation topic. I actually had a verbal encounter with the professor that was quite unpleasant because I felt the rug had been pulled out from under me, and I spoke out in a way that was uncharacteristic for me. I felt abandoned, but the EE Department stepped in and rescued me with continuing TA support and I found a wonderful advisor in Computer Engineering, James Gault, who more-or-less let me pursue my research without constraints (and without much intervention because it was not in his area of interest). I selected Jim Gault because he was kind and supportive.

Because of Jim Gault's background and expertise, I focused on Computer Engineering in my PhD studies, and my dissertation was on a micro and nanoprogrammable digital architecture that could interface between computers and other devices—networks, disks, channels, tape drives, and other devices. The generalized interface architecture in my thesis could be programmed to mimic any digital interface between devices. However the architecture I designed was too complex to be practical. As I began my research career at CMU I found that the language I had developed to program the universal interface took on a life of its own. The language, SLIDE (Structured Language for Interface DEscription) spurred my thinking about automatically generating specialized interface hardware from the desired functionality, functionality expressed using the SLIDE language. I then moved into design automation when I began my career as an Assistant Professor in Electrical Engineering at Carnegie Mellon University, specializing in an area we called synthesis—at first register-transfer synthesis and later high-level synthesis, to distinguish it from logic synthesis, which had its roots in creating gate-level digital hardware from Boolean expressions.

Again, a person was responsible for my move—actually two people. Victor Lessor, then a postdoctoral researcher at CMU Computer Science, said he would trade me his thesis, which I needed for my research, if I would send him my C. V. My C.V. ended up in the hands of the CMU EE Department chairman Angel Jordan. Jordan's wife was apparently the first female PhD in EE in Western Pennsylvania, and he was interested in recruiting me to CMU, a position I eagerly accepted because it was unusual coming from a lower-ranked institution like N.C. State to be hired as a tenure-track professor at CMU. Angel gave me a step up.

At CMU, I was interested in whether we could begin with an algorithm that described the functionality of special-purpose hardware or an instruction set that a digital processor was to execute, and then use software to automatically, autonomously synthesize the hardware, down to the microelectronic chip layout. Lower levels of the design process were being automated, so I focused on moving from an abstract functional specification to the register-transfer level hardware that implemented the function. We moved from using SLIDE to ISPS, a language for describing digital computer architecture based on ISP, invented by computing giants Gordon Bell (designer of many DEC computers) and Allan Newell, Professor and AI expert at CMU. First, we synthesized a small processor, the DEC (Digital Equipment Corporation) PDP-1. This opened the floodgates for my research agenda.

However, I always wanted to be in California. Furthermore, my natural demeanor, being helpful, meant that I began playing a role in the academic administration of the department and I did not know how to disentangle myself from the heavy service I had voluntarily taken on that was disproportionate for an assistant professor. Frankly, the word "no" was not in my vocabulary, and is still very hard for me to say. Also, in a bold innovative move, the department chair took funding destined for teaching assistants and converted it to research assistantship funding. It helped boost my research productivity significantly as I was able to quickly build a team without much external funding. It also trained me to diagnose and debug students' circuits, since I had no teaching assistants or graders for some of my courses at CMU and I gained valuable expertise that was useful later, as we will see. But it made the CMU slogan, "My heart is in the work" poignantly real as I worked more intensely than ever.

I was successful obtaining research funding from the Army Research Office, NIST and from NSF, thanks to collaboration on the NSF effort with a colleague at CMU, Ken Preston, and to my program monitor at the Army Research Office, Jimmie Suttle. Jimmie, although senior, had been pursuing his Ph.D. at NCSU while working full time in the Army Research office when I was also studying at NCSU. After he graduated, he served on my dissertation committee and even spent an afternoon helping me prepare my interview talk, a role that a dissertation advisor would normally take. Jimmie monitored my first synthesis research contracts, and was one of the major contributors to my success as a researcher. He was held in such high esteem by the communication science researchers at USC that they named one of their computers Jimmie, after him. He had encouraged me to make a move to USC because of the strength of the department, and his recommendation carried weight with then USC Executive Vice President and former EE Professor Zohrab Kaprilian, who recognized Jimmie's contributions to the EE Dept. Colleagues of Zohrab's used to say that I was hired at USC "in spite of" the fact that I was a woman.

So, I left CMU for the University of Southern California in the fall of 1980, where I felt welcomed and appreciated, although I found out later that the USC EE chair, George Bekey, had a tough sell to convince the "theory" faculty to hire me, as he documents in his book about the USC Viterbi School of Engineering[1] because I was an experimentalist (and maybe because of my gender?). In the meantime, I protected my time and energy, and fought my tendency to volunteer for every need the department had.

At USC, my students' software along with a commercial standard cell layout tool synthesized the chip layout for a JPEG image compression algorithm in 48 h. I was highly successful raising research funds, garnered an opportunity to have members of the USC Board of Trustees tour my lab, and my son was born in 1987.

My world changed after my son was born and my father passed away, in ways I could not have imagined, and I started searching for meaning in my work. In 1989, a well-known semiconductor device researcher and professor at Cal Tech, Carver Mead, published a book entitled Analog VLSI and Neural Systems, and I was hooked. I was so focused on current research and my family that this opportunity to

[1] A Remarkable Trajectory: From Humble Beginnings to Global Prominence, G. Bekey, et al.

make a change did not sink in for another few years, years I lost valuable time that I could have poured into the research, but I was busy with my current well-funded research into automatic synthesis of hardware, and with my young son. I was hesitant to change research directions boldly because the security of a funding stream and tenured position was strong motivation to stay in place. As it was, I did immediately start jotting down ideas in a lab notebook, and co-supervised a student pursuing a Master's degree in Computer Science, Charles Chao, with Michael Arbib, who had found his way to USC after I arrived. The circuits Chao built, under my supervision, kick-started the BioRC Biomimetic Real-Time Cortex project I launched in Jan., 2006. But I am getting ahead of myself. There is an old saying that if you miss an opportunity, you "let the brass ring go by," and someone else takes it. The brass ring was something you snatched off a stand next to a merry-go-round ride to win a prize, and if you did not take it the first time it came around, someone else would. There were no second chances. So the brass ring of artificial electronic neurons slid by almost unnoticed.

Bored with the design automation research that provided steady funding, I became an administrator. While my research into how to get software to design hardware proceeded, and I pushed forward to show how software could create network architectures for special-purpose systems, called Intranets, I felt a tug into university service because the new president at USC, Steve Sample, was a compelling individual and I wanted to be part of his administration. By then I was a Fellow of the IEEE and a full professor. I was discouraged by industry attempts to silence my efforts because I was in competition with commercial products. I was elected Faculty Senate President and served for a single day (my colleagues set up the one-day tenure as President so I could go on record as having served) and then I joined Sample's administration. I served for several years as Vice Provost for Research, and Dean/Vice Provost of Graduate Studies as well. After over three years in the USC Administration, I came back to Electrical Engineering, still searching for the research that would satisfy me. I thought that my future lay in higher-education administration, but it did not work out that way. I began searching for something interesting and/or compelling to focus my research agenda.

During my hiatus from the intense research activity I had performed before entering the USC administration, I carried out a study for the EE Department (Chair Mel Breuer) on whether departmental hiring should focus more on nanotechnologies and other advanced technologies, and gained not only summer salary but also valuable expertise that would prove useful later when I began research into artificial neurons. I studied carbon nanotubes as a likely technology for implementation, being influenced greatly by the research Charlie Lieber was performing at Harvard, with nanotubes and living neurons. Like much of what I have learned in my life that proved to be useful in my career, it was initially learned "on the job" when I needed extra income. My short exploration of nanotechnology was instrumental in shaping the research to follow.

It was after being involved in a couple of the DARPA Grand Challenges in 2003 and 2005 to build an autonomous vehicle, I was struck by how much better our jeep might have performed in 2005 if the code had been more brain like. I also noticed the only articles and books I was reading were about the brain. I could not force myself to read even the articles that described research based on my own research or that referenced my work because I was finished with anything not brain-related. Cold-turkey, full stop, I dropped everything related to hardware synthesis research (all relevant students had graduated). The next January, 2006, I launched the BioRC Biomimetic Real-time Cortex project, with the goal of fundamental research into how to build a system that mimicked human intelligence. I had some modest goals at the advanced age of 58 to reinvent myself as a neuromorphic engineer, to do some basic research to build the foundations of artificial brain electronics pre-retirement. However, again, my professional life changed dramatically. To my surprise as a newcomer to the neuromorphic field, the handful of students I had recruited onto the BioRC project and I started producing results!

12.4 The Real-Time Cortex

One of the critical questions in modeling the brain with electronics is what mechanisms and details must be included to achieve intelligence and reasoning capability. We included mechanisms in our electronics that others did not, and showed that we could achieve some desired behavior as a result. This was not a proof that those mechanisms were required. It was experimental evidence, and we continue that line of reasoning to this day.

12.4.1 Nanotechnology

I began the brain research by believing that complex computations performed by individual neurons were the key to building an artificial brain, and that nanotechnologies were the solutions to solving this problem. I was fascinated by the dendritic computations proposed by John Elias from the Cal Tech group, and thought we could easily build similar structures using MOS/nanoelectronic transistors. By 2007, I had obtained NSF research funding for the BioRC project. I had been particularly fascinated by the promise of nanotechnology as a vehicle for an artificial brain, given the possibilities of extraordinarily small transistors, fabrication of circuits in three dimensions, and circuits that could reconfigure themselves as brain connectivity needs changed. My earlier experience assisting the department with hiring planning in nanotechnology became critically useful.

I have always been ahead of the curve in my research agenda, with a paper published in 1984 on image understanding still being cited as I write this in 2019. In 2007, nanotechnology was just beginning to be developed in ways that could support electronic circuit investigations. By 2008 the best we could do with nanotechnology was to fabricate a crude electronic synapse out of a handful of resistors and a single transistor whose channel was formed with parallel carbon nanotubes, using

Chongwu Zhou's nanolab and his CNT transistor. Chongwu's student Jialu Zhang and my student Jon Joshi were not able to get the synapse to behave as a biological synapse, so I marched myself into the nanolab, commandeering the two students for necessary expertise, because I had never used the nanolab equipment we needed to demonstrate a working circuit and I needed their help. A bit of fiddling with the equipment and an afternoon spent selecting testing parameters and circuit components (resistors and capacitors) resulted in our circuit with a nanotube transistor behaving like a synapse. My observation that led to this result was that the test bench itself introduced significant capacitance into the circuit and that we needed to slow down the input waveform used to model a spike significantly to overcome the capacitive effects of the test bench. Again, my experience from CMU days teaching undergraduates to construct circuits in the laboratory, and debugging their non-working circuits allowed me to diagnose and repair our stubborn synapse and coax it into mimicking a biological synapse.

As I explored neuroscience further, I became convinced that connectivity between neurons was a difficult problem not solved by connecting neurons together in a shared network, called Address Event Representation, in a manner that did not address the exquisite timing of brain signals that could not be delayed or their timing adjusted arbitrarily, as occurs with shared communication. While current neuromorphic systems adjust the timescale so that shared, networked communications are "fast enough," there remain nuances of timing that are unexplored. Synchronization of spiking neurons in the biological brain appears to be a significant event that is associated with focus, concentration, and awareness. Nanotechnologies might be able to provide connectivity without the timing awkwardness of shared networked connections.

It turned out there was some interest in what we were attempting with nanotechnology, so I gathered a larger group of PhD students, most of whom were floating in my department at USC without an advisor or direction, and supplemented the team with MS students eager for research experience. We set about showing how to build a brain, not based on others' engineering research except the seminal book by Mead *Analog VLSI and Neural Systems* and the research being performed by former Cal Tech students of Mead's. Early on, I stumbled on a paperback *On Intelligence* by Jeff Hawkins, who had designed the Palm, an early hand-held device, and had set his current direction on solving problems using software modeled on brain characteristics. His book confirmed what I had thought were two major engineering challenges for artificial brain construction, scale, and connectivity. My colleague at USC, Bartlett Mel, spent time with me explaining the exquisite specialized nonlinear computations that each biological neuron performs. Hawkins' and Mel's understandings of the human brain led me to three major problems that had to be solved for an artificial brain to become a reality: the nonlinear complexity of computations within each neuron, the scale of brain at 100 billion neurons and trillions of synapses that connect neurons to each other and the dense connectivity between neurons that dwarfed the connectivity capabilities of our modern CMOS circuits with less than a dozen layers of metal for each integrated circuit. Each neuron in the human cortex is signaled by up to 100,000 others, and each neuron sends its signal to 10,000 others. It appeared that nanotechnology of some sort could be used because of the obvious small size, the capabilities of complex transistors with multiple gates, and the possibility of growing nanotechnologies in three dimensions, supporting great connectivity. I focused on studying neuroscience, not engineering. Complex nonlinear dendritic computations were the subject of one dissertation from my group that simulated neurons constructed from carbon nanotube transistors. This dissertation showed that nonlinear dendritic computations could be used to model individual neurons that performed complex functions, like border ownership, determining which side of an edge in an image was an object and which was background.

We began to study the retina (considered a part of the brain), and found the physiology of retinal computations to be different from the rest of the brain. Most neurons in the retina do not fire, for example. Although the retina had been the subject of the earliest contributions to electronic neurons by Misha Mahowald, she passed away, and her research was continued by others. It has not attracted the attention of the neuromorphic community it deserves. Without the retinal processing performed by the biological eye, the information processed by a neuromorphic brain is raw and unfiltered, leaving the visual cortex to do the heavy lifting. With valuable inputs from Norberto Grzywacz in Biomedical Engineering, we produced a dissertation on electronic retinal circuits, showing how they convey edge motion to the brain as well as perform more basic functions like contrast enhancement and adaptation to overall brightness.

As my understanding of the brain construction problem increased, the show-stopper and unsolved problem seemed to be the enormous plasticity of the human brain. Not only do synapses strengthen and weaken in milliseconds, but synapses, connections between neurons, form and dissolve, albeit more slowly. Because nanodevices and nanowires, unlike conventional integrated circuits, could potentially assemble and disassemble during operation, once again the technology candidate for neural electronics seemed to be some form of nanotechnology. One dissertation demonstrated STDP (Spike Timing Dependent Plasticity, a form of synapse strengthening and weakening) and presented an electronic model of neural remapping called synapse borrowing, where loss of sensory information remapped existing quiescent neurons to be remapped (reconnected) to process other sensory inputs. Again, collaboration with Tansu Celikel in neuroscience led to our example electronic model focused on the barrel cortex of the rat with whisker loss but the circuits were more general, and was a major contributor to that dissertation by Joshi. We also electronically modeled a little microscopic worm with 302 neurons, *C. elegance*.

12.4.2 Astrocytes, the Neural Code and Brain Repair

As I dove deeper into the neuroscience literature, I fell in love with astrocytes, cells in the brain that are a type of glial (glue) cell. They are pervasive and contribute to computation as well as influence blood flow and contribute nutrition. Their interaction with neurons can convey activity between neighboring neurons, encouraging firing and removing excess neurotransmitters when activity is too intense. Astrocytes have been shown to synchronize neural firing, and to preserve connectivity between neurons when individual synapses fail, maintaining a form of homeostasis in the brain. The presence of astrocytes challenged us to enhance our neural designs. Electronics modeling astrocytes in conjunction with neurons have formed the basis for two of the doctoral dissertations emerging from the research.[2]

The figures here (Figs. 12.1, 12.2, 12.3, and 12.4) show an example ocular network in the brain, with astrocytic monitoring of the neurons processing the left and right eyes. We were able to show that when a synapse on *N1* failed to transmit a signal from the left eye, the astrocyte stimulated a synapse on a different neuron *N2* to begin transmitting that signal. The shapes of the spikes emerging from the surviving neuron indicate whether the signal is coming from the right or left eye.

The waveforms in Fig. 12.2 show initially that both *N1* and *N2* are functioning normally and outputting *POSt1* and *POST2* spikes. At about 200 µs (the signal would be much slower in biological tissue), the dendrite processing the left eye exciting neuron *N1* ceases operation, and *N1* ceases spiking. The astrocyte calcium ASTRO_CA decreases, and at about 360 µs, the astrocyte releases glutamate GT_S4, a transmitter that excites synapse *S4* to take over processing for the left eye. By 380 µs, *N2* starts spiking with a different spike shape signaling the left eye signal.

Zoomed-in portions of the waveforms for *AP_POST1* and *AP_POST2* for a similar experiment are shown in Fig. 12.3 below. The top two waveforms are for normal processing of the left and right eyes. The bottom waveform shows *N2* processing both eyes with different spike shapes.

The final ocular network is shown in Fig. 12.4.

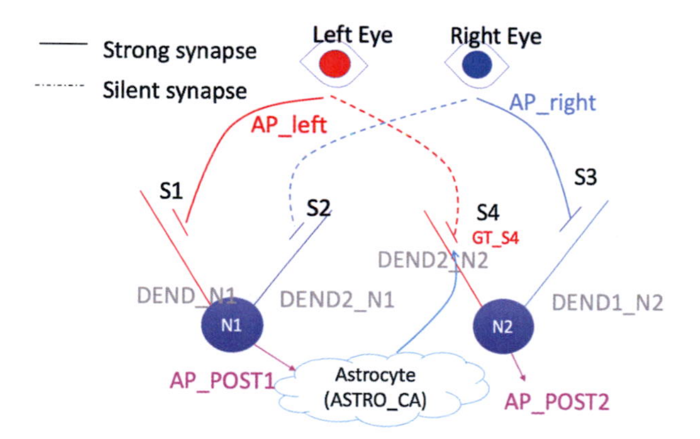

Fig. 12.1 Example ocular network with left and right eye signals (Source: Internal research documents, and Rebecca Lee's dissertation presentation)

[2] It should be noted that three of the Ph.D. students on the BioRC project (about 1/3 of the Ph.D. students) were women, Chih-Chieh Hsu, Rebecca Lee, and Yilda Irizarry-Valle, and their contributions were key and essential to the results described here.

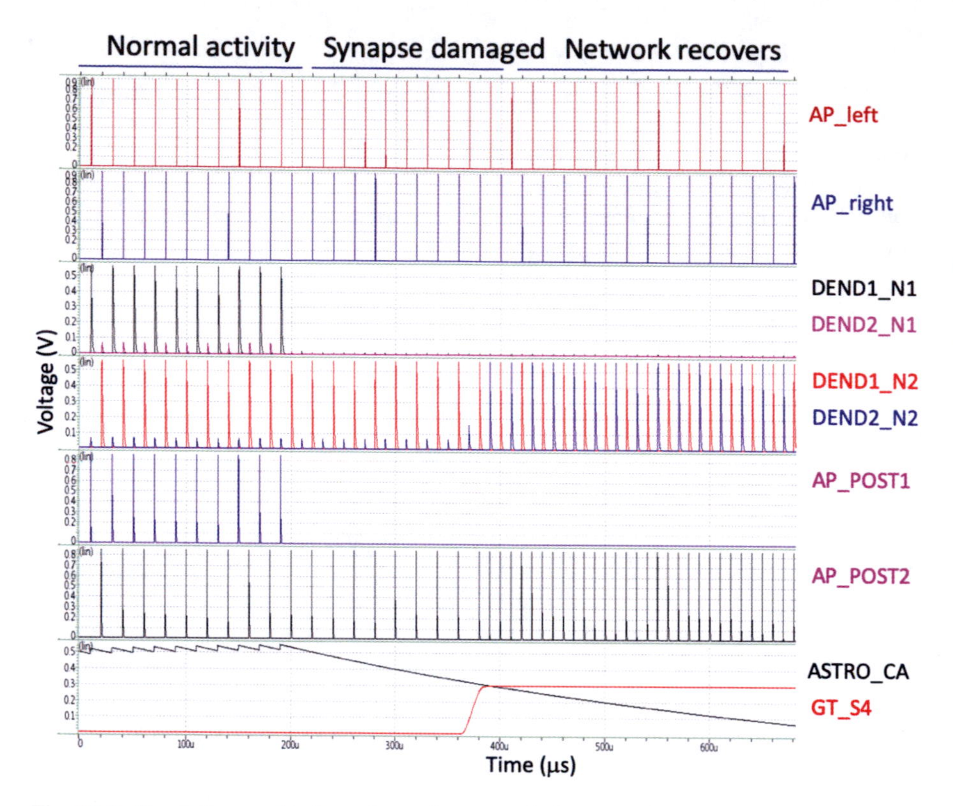

Fig. 12.2 Simulation results from the ocular network demonstrating that network can recover from loss of a neuron

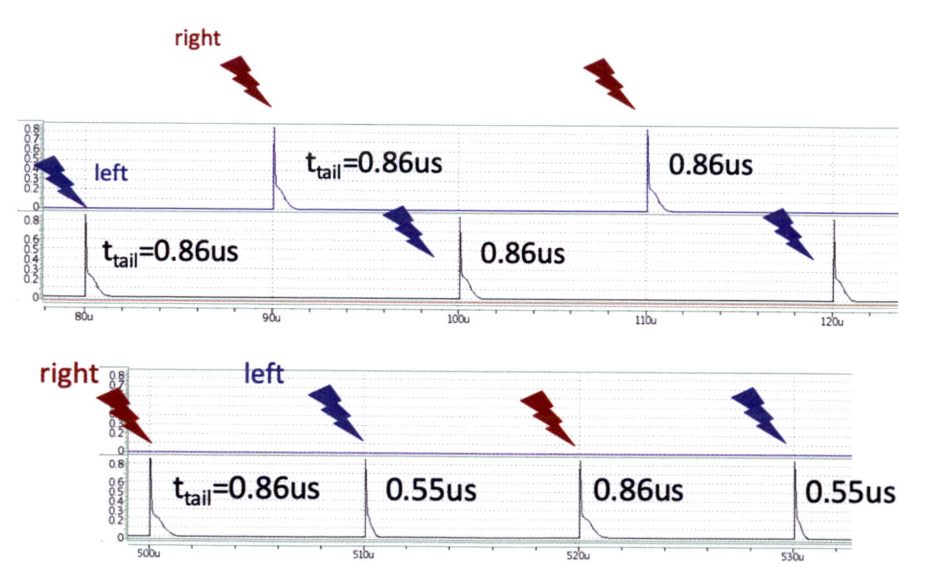

Fig. 12.3 *AP_POST1* is shown in the upper, blue trace, and *AP_POST2* is shown in the middle, black trace. Durations of the repolarizing tails are also shown. In the bottom trace, where *N2* is processing both eyes, it can be seen that the repolarizing tails are different for traces belonging to the left and right eyes

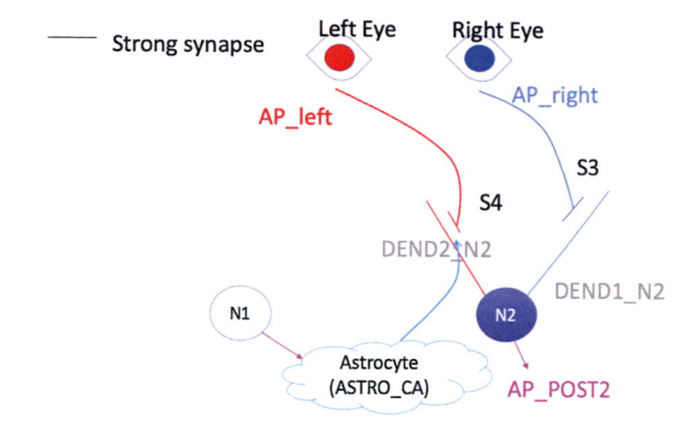

Fig. 12.4 Ocular network remapping after recovery from damage in N1. Note that neuron N2 is now processing both left and right eyes

12.4.3 Current Status and Future Plans

Our library of neural electronics circuits has been complete enough to model schizophrenic hallucinations, OCD, and other brain disorders. In addition, we have shown how dopamine strengthens synapses as a form of reward during learning, without supervised training. All these capabilities were demonstrated in a dissertation. In another dissertation, we have shown electronic models of neurons that behave in a variable fashion, either noisy or chaotic. A dissertation just completed presents redesigns of the electronic circuitry within individual neurons resulting in ultra-low power consumption. We have simulated electronic circuits using several forms of nanotechnology that form the basis for another dissertation in preparation. While we could continue to perfect our neural circuitry, I am now focused on showing how learning can progress without forgetting in our electronic neurons, in collaboration with Francisco Valero-Cuevas and his robotic cat Cleo, which is a NeuroBot, a robot with a nervous system.[3]

As a final challenge, I am grappling with the neural code that underlies how information is transmitted between neurons as well as how DNA affects neural behavior. We have shown neurons with different spike shapes indicating the origin of the excitation, and neurons spiking with different frequencies to indicate different meanings. We are pushing forward to examine more-complex coding and the entropy of brain signals. Somewhere in this neural code there might be clues to understanding consciousness, but that could be decades away. The choice to retire and sit on the beach surrounded by silicon crystals (sand) seems boring compared to the obvious choice to keep using silicon (and other materials) to build a brain.

I would like to acknowledge department chairs and colleagues over the years for their support and intervention when it mattered. My students are like a family to me,

[3]All my recent publications and my C.V. can be found on my research web page http://ceng.usc.edu/~parker/BioRC_research.html.

and from undergraduate students to postdoctoral researchers, they have been invaluable in contributing to the success of our research. The WiSE program[4] at USC gave me confidence that someone who cared, someone who had my back was there. My son Joe Bebel deserves more than acknowledgement for his incredible help in my ability to function in this increasingly technological world and for his support.

I am very proud of my ability to move between fields, to create research areas that had not existed or only in small efforts, and to learn a new field (neuroscience) at age 58. As I mentioned before, three of the Ph.D. graduates of my BioRC project are women. Women are survivors and women know how to make something out of almost nothing. I feel very uncomfortable describing my strengths, and that sense dogs women as they move to succeed in STEM. However, I have proven my capabilities, and the evidence exists. I am not an imposter.

[4]Women in Science and Engineering.

Chapter 13
VLSI and Beyond: The Dream of Impact, Creating Technology with Inclusive Cultures

Telle Whitney

13.1 The Beginnings: My Childhood

For me, life started in Salt Lake City, Utah. Both of my parents had deep roots in Salt Lake City, with Mormon heritage going back many generations—my father was descended from Brigham Young. My father, an only child, married my mother when they were still in college. She supported him through college and law school at the University of Michigan. They spent time in Guam while my father was in the service. My mother's dream was to be married and to have a family, and my father's dream was to get out of his home. I was the oldest of three children. I was a happy child, and my parents were very loving and encouraging, and led me to believe that I could do anything, Fig. 13.1 is an early picture.

The disconnect in my parents' marriage came to a head when I was seven, and they separated and divorced. My father quickly became involved with his second wife Beverly, who was the love of his life. Beverly had three kids, and together they had a child. Although my mother's history was in Utah, she was raised in Los Angeles, and she returned to her roots, and to where her sisters lived, in Southern California. Thus I grew up in a household of four women, my mother and two sisters, in the San Fernando Valley. Although economically we were challenged—my father supported us in addition to his new wife and family, I never thought of us as poor. We got by, and our world was very middle class—lawyers, doctors, businessmen. My mother believed in me, she taught me that I was special. I believed I could do anything, and was very unaware that most of the areas I was drawn to were jobs of men. I distinctly remember the 1969 moon landing and being completely captured by the technology and what was possible.

But life got tough when my mother was diagnosed with Multiple Myeloma. It was a death sentence, although I did not know that. She had finally received her

T. Whitney (✉)
Co-founder, Grace Hopper Celebration, Former CEO, Anita Borg Instiutte,
Palo Alto, California, USA

A. C. Parker, L. Lunardi (eds.), *Women in Microelectronics*, Women in Engineering and Science, https://doi.org/10.1007/978-3-030-46377-9_13

Fig. 13.1 Telle Whitney as a child

degree in history, and was teaching. She was even beginning to date. Although my mother continued to have incidents, and kept going to the hospital, she was the center of my life. My best friend Abbie lived nearby and her mother was my mother's best friend. My life was connected closely—my mother, my sisters, my mother's family, and my best friend. And I thrived—I was great in school without trying, I skipped a grade, I rebelled against what was expected —but I thrived. Then the bottom fell out when my mother died—I was 15 and had just finished my sophomore year in high school. In hindsight, we were very lucky that my father had always stayed connected, but it certainly did not feel like luck. The three of us moved in with my father and his family in Salt Lake City. In the blended family, there was the California girls, Beverly's children and the youngest, which was both of theirs. I hated it there, I hated my stepmother, and I hated life.

I was determined to be self-sufficient and I was smart. Because of my experience with my parents, I was NEVER going to be dependent financially on a man—that was my driving force. But I also did not ask for help or guidance. Although I was great in math, I stopped taking it after my junior year because I did not believe it was good for anything. I became involved with theater.

I arrived at the University of Utah and majored in Theater. The University is a large public university but it was respected. I started in theater but quickly found that was not where I belonged. I considered Political Science, and then I came close to dropping out. My stepmother had suggested an interest inventory test to look at what fields I would be drawn. Because it was her suggestion, and she could do not right, I ignored the advice until I was out of options. I finally took the test, and computer programming came out way ahead of any other field. I took a programming class—COBOL—and fell in love with the field. University of Utah was one of the early Computer Science departments, and an early member of the Arpanet. There were women around—the COBOL class was taught by a woman, and there were female graduate students. Although the classes themselves included mostly men, I was so thrilled to love what I was studying, that I hardly noticed. I was behind because it took me over a year to find the major. I focused on the classes and absorbing the knowledge as if I was parched and had finally found water. I could not learn enough or fast enough to keep up with my thirst for knowledge. It took me five years to complete the degree while I was putting myself through school by waitressing. I did not have a car, I kept my expenses light, and I had Social Security from my mother's death. I lived in an apartment near campus, and I survived, and my life revolved around school. I had light support from my father and stepmother, but mostly I was determined to be self-sufficient.

13.2 Computer Science: My Foundation

I immersed myself in the Computer Science curriculum, with late nights, towering decks of punched cards to submit for the programs, and learning many varied programming languages—Pascal, Snobol, Lisp, and many others. I had friends who were interested in Architecture (the kind where you design buildings), it sparked my interest, and I took an elective in Computer Graphics. University of Utah was an important academic location for Computer Graphics, home of the company Evans and Sutherland, and producing many famous professionals in Computer Graphics including Ed Catmull. But most of the instrumental graphics faculty had moved on. The instructor of the Graphics class was a faculty member Rich Riesenfeld, whose wife Elaine Cohen was also on the faculty, they were both very involved with Computer Aided Geometric Design, a much more mathematical intensive field. Rich taught the class, and he noticed my work. We had a visiting lecturer to the class—a man named Ivan Sutherland, who was a graphics legend. Rich introduced me to Ivan who had co-founded the Computer Science department at Caltech and Ivan invited me to apply. I was definitely out of my comfort zone, I had never seriously considered graduate school, let alone a PhD. But, as scared as I was about the entire opportunity, I could not imagine passing it up. I applied. My grades were not great from my early days, but I had straight A's in the CS classes I took. Rich gave me an outstanding reference letter, and arranged for a colleague to give me another letter and I was accepted. My boyfriend at the time was not at all happy about the idea of me attending Caltech, and leaving Utah. I was burned out after five years of undergraduate school and working so I postponed leaving. I took a job at Sperry Univac in Salt Lake City creating an entrepreneurial product for the Israeli army. The team for the product developed the operating system, as well as the application on a bare bones Univac machine. Whereas the idea to develop everything from scratch seems silly in hindsight, for a new college graduate it was a great experience.

We wrote in assembly code, and I had the chance to develop a debugger using a bare machine with hardware switches. The yearlong engagement with Univac gave me an understanding of how computers really work. But I did not forget about Caltech, my boyfriend could not come to grips with my plans and left, and I moved forward. I applied again, and they accepted me. It was an important lesson to me to *Take Risks when opportunities arise, and go for it.*

I arrived at Caltech in the fall of 1979. It was an incredibly exciting time at Caltech. Ivan Sutherland had co-founded the department with a Caltech professor Carver Mead. Ivan and Carver had recently published an article in Scientific American on the possibilities of VLSI [1]. A pre-print of the first few chapters of the Mead and Conway [2] textbook was available for the VLSI course I took during this first year. It was an exciting time, and like many first year students, I was completely immersed in an overabundance of required courses, including the VLSI course. My education at the University of Utah had not prepared me adequately for the rigors of Caltech. It did not help that it had taken me time to find Computer Science as an undergraduate. I spent the first year at Caltech scared to death that I would be found out, what is often called the Imposter Syndrome. I worked really hard and I did my best, but I walked around campus with a knot in my stomach sure that I would be found out, that I did not really belong as I compared myself to the other talented first year students from prestigious universities such as Berkeley, National Taiwan University, and University of Illinois, Urbana-Champaign. *But the resilience I had learned from my difficult childhood served me well, and I kept showing up every day, doing the best I could.*

In that first year, I had very little exposure to other woman. There was another woman graduate student who was further along, but she did not appreciate other women. This was one of the rare times when I saw women not helping other women, my experience was that the few women who were more senior than me helped me out in any way they could. At the end of the first year, we had an exam called mini-orals. These were a shorter version of the official doctoral orals, but graduate students regularly failed the mini-orals, and it was common knowledge that if you passed the mini-orals, you could stay. I passed, and remarkably, I passed easily. I headed off to Digital Equipment Corporation for the summer, with the knowledge that I would remain at Caltech.

13.3 Navigating VLSI

I attended Caltech to work with Ivan Sutherland. He put together a small team of visiting scholars and industry people, and we met regularly, exploring tools for supporting increased chip complexity as well as models for circuits. My master's thesis topic was *A Hierarchical Design Rule Checker* [3]. Much of the focus of the VLSI work in those days was based on the theory that you could design complex chips by creating regular structures of simpler elements. Before you sent a chip design for fabrication (often called taping out because of the early process of creating a tape

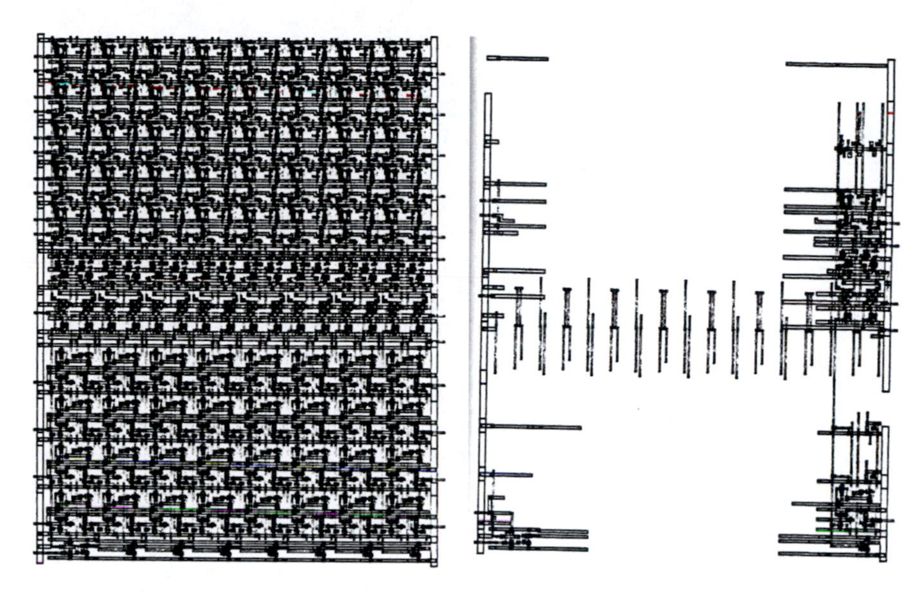

Fig. 13.2 LRU and filtered LRU CAM

with small geometrical elements that form the chip), it was important to ensure that the chip met a set of design rules, and costly if design rule violations created a non-working chip. These rules embodied electrical rules dictated by the fabrication process. As chips became increasingly complex, we were looking for a way to simplify the amount of information that we needed to check. As an example, Fig. 13.2 includes a copy of a Least Recently Used Content Addressable Memory (LRU CAM) in its fully instantiated state, and filtered to show unique portions of the design. It was an exciting time, and for a young student, provided the chance to work with extraordinary people. I particularly remember an industry review meeting that happened two months after I joined Caltech. Ivan informed me I was to give a presentation (oh my goodness, I cannot possibly do this). The presentation was short, and I did what I could to prepare. Afterwards, Ivan took me aside and told me all the things that I had not done right, but also congratulated me on the success of the presentation. I presented, even though I was scared, and I learned.

After my first year at Caltech, Ivan let me know that he was leaving Caltech and reminded me that he had not made a commitment to advise me during my PhD. I was very disappointed, but realized over time that it made more sense to work with a faculty member who remained at Caltech. I spent the following year writing up my MS thesis as well as taking additional courses. When I finally sent my thesis to Ivan, he responded that he did not like passive verbs and I needed to re-write the thesis using active language. It felt like finishing my MS thesis would never end. Eventually he signed off and I entered my next phase.

Carver Mead became department chair at Caltech when Ivan left. He approached me with rudimentary ideas he had for a circuit level representation. We agreed to work together, and thus began one of the most important partnerships of my life.

Those early conversations about circuit level representation became the *Pooh* representation [4], supported by an interactive graphics system called *Tigger.* Fig. 13.3 illustrates a Pooh representation for a simple bit register. Fig. 13.4 shows a simple register in the Tigger interactive system. Although this representation provides any angle chip layout, it also embodied the circuit description. In practice, we could use an integer based grid and restrict the geometry to 45 degrees.

At this time Silicon Compilers were an important research area, creating layout and circuits from a high level specification. Dave Johannsen had developed a Silicon Compiler a few years earlier [5]. His software provided the ability to generate a complete chip design from a high level specification, if you could map the specification to a parameterized datapath. Each year, Caltech hosted industry representative that spent a year at Caltech. In the 1981–1982 year, I worked closely with a number of these representatives on a next generation Silicon Computer [6] using the Pooh circuit level representation as its building blocks. We created an embedded language interface, developed a Datapath Generator, memory generator, and a global router that would allow complex chips to be created. By using Pooh as the underlying representation, the software maintained the circuit level connectivity as well as layout. This was an exciting time and what was great about this period was working with a lot of smart people and big ideas. At the time I also worked for my first startup company Silicon Compilers, founded by Dave Johannsen and others. John Doerr was the company's first CEO. It gave me a taste of entrepreneurship, although my engagement as a part time employee was short.

Although I spent the next few years working on my VLSI based PhD topic, Carver had moved on to a new field. He created a new area of research based on neural nets. Carver was a silicon device guy and he wanted to use the analog properties inherent in silicon to model the neural nets in our brain. I moved with Carver to his new analog lab, but I continued on a digital topic. Once again, the exciting part of this period was the chance to be around intellectual giants including Carver. Carver, John Hopfield, and Richard Feynman offered a class on neural nets in those years as they explored ways to bring Computer Science, Engineering, Biology, and Physics together to create these neural net analog systems (https://en.wikipedia.org/wiki/Computation_and_Neural_Systems). I spent a total of four years completing my thesis, and luxuriated in an environment that encouraged experimentation and breaking new ground.

What was also important to me was my relationship with Carver. By this time, Carver was famous, but he always made time for his students. He created an environment of learning and he was always available to me to discuss ideas. I have since learned the importance of mentoring, but the person who demonstrated to me what mentoring could be was Carver.

There were also a number of extraordinary women in Carver's lab during this time. They include Marina Chen, Lounette Dyer, and Misha Mahowald. Although Caltech was predominately led by men, and the student body was 13% women, I had a few brilliant women around me.

Fig. 13.3 Illustrates a Pooh representation for a simple bit register

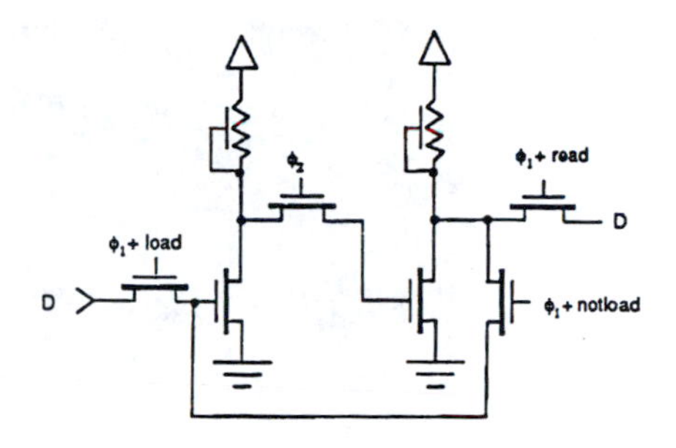

circuit diagram

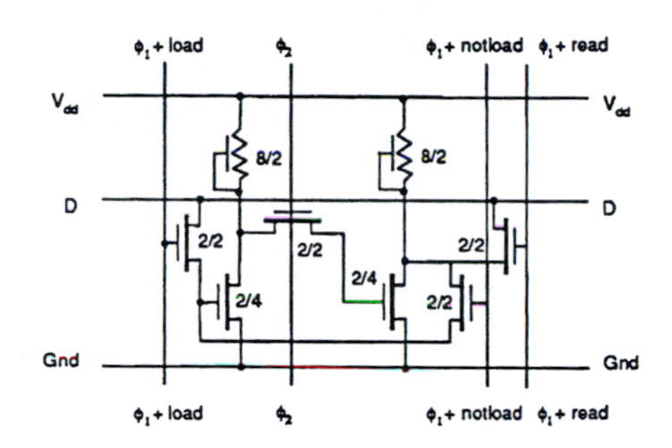

topological sized schematic

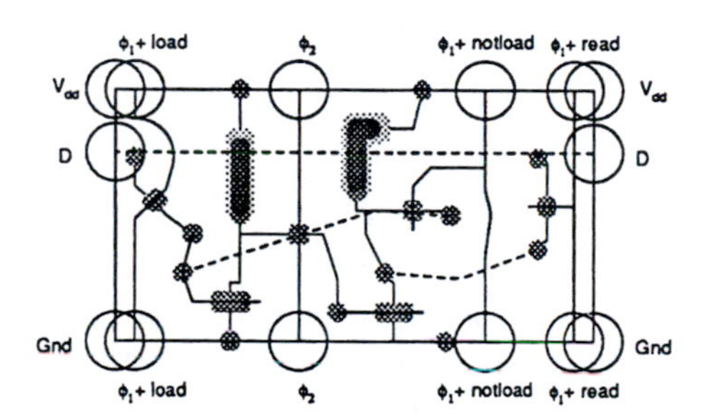

stick diagram

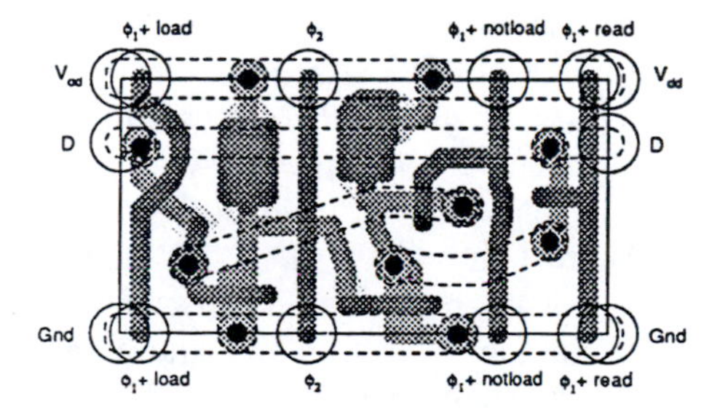

mask geometry

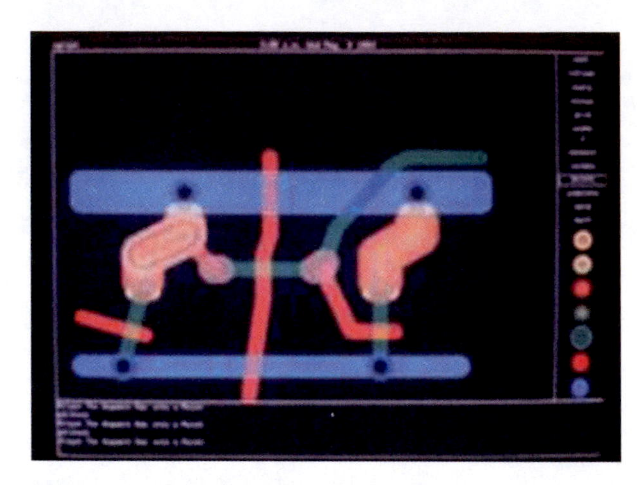

Fig. 13.4 Shows a simple register in the Tigger interactive system

13.4 Entrepreneurship in Silicon Valley

After graduating from Caltech in 1985, I arrived in Silicon Valley in early 1986. I spent a lot of time interviewing and deciding on what was next after graduation, ultimately I went to Schlumberger Palo Alto Research center (SPAR). But Schlumberger decided shortly thereafter that they did not want to be in the Computer Science Research area, and I left after 18 months. It was a lesson that you can review and process a lot of data when making a decision, but outside and unforeseen factors often influence the outcome, in other words, *control is an illusion*.

But I also realized that I was born to be an entrepreneur. I joined a Field Programmable Gate Array (FPGA) company called Actel as employee number 30. Actel was founded by a former student of Carver's—Amr Mohsen. There were a number of FPGA companies at this time, and the vision was that a customer could create a specialized design instantly by programming a reconfigurable chip. This approach compared to fabricating a specialized chip, which of course takes months or even years. The Actel technology used an Antifuse technology. An Antifuse is small, roughly the size of a transistor, and is programmed or blown by applying a high voltage. Once an Antifuse is programmed, the programming is permanent.

My role at Actel was to help programmatically generate much of the chips. We applied many of the Silicon Compiler techniques that we used at Caltech. I developed an embedded language system, and used software to generate much of the routing channels that included regular placement of Antifuses to allow programmable routing.

I grew up, professionally speaking, at Actel. I was very involved with the chip design of the Actel 1280, the largest and densest FPGA of its time. I led the Act3 feasibility study, providing me with a first taste of leadership. I led a project that overhauled the customer software, designing it to meet current software and usability standards. Finally, as a software director, I led the customer focused logic synthesis and timing efforts. There was very interesting work going on at that time to automatically map a high level logic specification to FPGAs. The best software incorporated both logic synthesis with layout knowledge to ensure adequate timing. We were able to create very accurate timing models based on the silicon, and then optimize the logic synthesis based on these timing models. It was also increasingly common to have software IP modules that were optimized for the FPGA, and could be used during the synthesis process.

But my entrepreneurial bug called me and I accepted a position as VP of Engineering at a company called Malleable Technologies. I was employee number four, and a member of the founding executive management team. The Malleable founder, Curtis Abbott, I met through Carver Mead.

Over the next 18 months, I went on a start-up roller coaster ride. Our technology, envisioned by Curtis, was a programmable architecture. It had a simple but Digital Signal Processing (DSP) calculation rich instruction set. In addition to memory, the chip included a programmable logic array. Although the chip was programmable, the product we developed used the programmable nature to support the product as a Voice Over IP processor. As VP of Engineering, I was responsible for delivering a high quality chip design, providing product engineering for this chip, as well as customer software to use the product. We also created a world class DSP algorithms team. We delivered a high quality functioning product on time. Subsequently we were purchased by PMC-Sierra to become part of their DSP product offerings.

It was an exhilarating time. I am definitely happiest as an entrepreneur. But being part of PMC-Sierra did not work for me, and I left. Subsequently the bottom fell out of the DSP market in 2001, and this product as well as many other products, never launched successfully. The Malleable product required both hardware and software to succeed, and most of the PMC-Sierra products did not have a software component, and ultimately it was a mismatch.

I also had a strong foundation in my personal life. I was happily married, to someone who supported me in what I wanted to do. I had built strong relationships with both my father and stepmother. She was not as terrible as I had believed as a young adult, in fact she was instrumental in creating a strong blended family. It was the strength of my marriage, as well as my deep friendship with close friends that allowed me to thrive in the work that I loved.

One other observation is the importance of Carver in my journey. In the previous section I discussed Carver Mead as an important mentor. But he also made introductions that led to my first three jobs in Silicon Valley, a role that is often called a Sponsor. A mentor is someone who can provide advice especially at critical junctures in your career. A sponsor is someone who recommends you for important positions that you do not even know exist. Sponsors are often critical in large organizations, where a successful career requires you to accept new and challenging positions. But as someone who regularly worked in small organizations, Carver became an important sponsor, introducing me to exciting new opportunities. For anyone who wants to become a leader, *finding the right mentor and sponsor is critical.*

13.5 The Big Transition: From Technologist Entrepreneur to Non-Profit Leadership and Entrepreneur

When I arrived in Silicon Valley, excited to be part of the great Silicon Valley entrepreneur future, I was also very committed to finding and connecting with other women. During my first year of living in Palo Alto, I connected with a group of women who had graduated from Stanford in Computer Science. It was through this group, many of whom remain friends today, that I met Anita Borg. Anita had graduated from NYU in Computer Science, had worked in Fault Tolerant Computing, and was a researcher at Digital Equipment Corporation's Western Research Lab (DEC WRL). She became one of my closest friends and an important partner in my future. While I navigated the Semiconductor world, with exciting technology but very few women, Anita and our other friends help keep me sane.

It was during our first few years of knowing each other, that Anita and I brainstormed on what we could do to both highlight the contributions of women, and help women stay in the field. One of the first persistent efforts was when Anita and 12 other women working in Computer Systems met at the Symposium on Operating Systems Principles (SOSP) computer conference and agreed to stay in touch. This agreement ultimately led to mailing list called Systers, which became one of the first broad based online communities, focused on women.

Anita and I spent many hours discussing the needs of technical women, and why it was important to have a safe space. Systers was initially inclusive of women working in Systems, but as it caught fire, women who self-identified as working in Computing were invited.

Based on our many discussions, and catalyzed by work of the Computer Research Association's committee on the advancement of women (CRA-W), Anita and I decided to create a conference that celebrated the achievements of women, and to name it in honor of one of the earliest Computing icons—Grace Hopper, who had recently passed away. We held the first Grace Hopper Celebration of Women in Computing Conference in June 1994, in Washington DC. Anita and I created all aspects of the conference. Neither one of us had created a conference, but as an entrepreneur, I always believed that anything was possible. CRA agreed to be the fiscal sponsor of the conference. Every woman we asked to speak at the conference said yes! The National Science Foundation (NSF) provided student scholarship money, and several companies agreed to sponsor the conference. As one example of the caliber of the speakers—there are three women who have received the Turing award as of 2019 (the Nobel Prize of the computing community). All three of them were scheduled to speak at the 1994 Grace Hopper Celebration—Fran Allen, Barbara Liskov, and Shafi Goldwasser (who ultimately was unable to travel to attend the conference). There were 500 people at the conference, including a few men. It was a remarkable success, and it broke even financially.

Over the next few years, we continued to hold the Grace Hopper Conference every three years. In 1997, Anita founded the Institute for Women and Technology. At that time I was deeply immersed at Malleable Technology. When the Institute was in an early stage of development, Anita was diagnosed with brain cancer. In early 2002, I left Malleable, and was consulting. I agreed to help Anita raise money. The Institute's board had embarked on a search for a new CEO, but the organization was not in financial shape to attract a CEO. I reluctantly agreed to become the interim CEO, as we figured out the next steps. Because of Anita's illness, the organization was in difficult shape. I did a lot of soul searching about what was important to me. The organization had an incredible board including Maria Klawe, Bill Wulf, Greg Papadopoulos, Rick Rashid, and Linda Bernardi. I had their support, and the chance to make a difference in thousands of people's life.

I became the CEO of the Institute for Women and Technology in June 2002, renamed the Anita Borg Institute after Anita's passing in 2003. Accepting this position definitely was a decision to step away from my technical roots, and to focus on entrepreneurship and leadership. We had to reset the offerings of the Anita Borg Institute, and over the next 15 years I led the organization to one of global prominence. The 2017 Grace Hopper conference included over 22 K attendees, 95% of them women. Figure 13.5 is a picture of me at the conference. The conference attendees were between one third and one half students. The organization, through its key programs, reached over 750 K people. We created an industry partner program, and championed the once-obscure concept of driving business value through gender diversity. We created programs such as the Top Companies program that provides tools to organizations to address their diversity and inclusion commitments, because what you measure you will change. In my position as head of the Anita Borg Institute, I regularly had young women approaching me letting me know

Fig. 13.5 Telle Whitney at the Grace Hopper conference

that the conference changed their lives. We also saw the rise of many complementary non-profits that worked with women and under-represented minorities, many of these organization worked with us, or learned from us. The position was the most satisfying position of my career. One of my life lessons was the importance of considering ninety degree turns when the opportunities arise, and remain open to the possibilities. *Many careers have multiple acts.*

I also had the good fortune to meet Lucy Sanders and Bobby Schnabel, both from the University of Colorado, in my first year as Anita Borg Institute CEO. They had ideas about providing resources for organization across the pipeline—K-12 through senior academic and industry leaders. We joined together and founded the National Center for Women and IT (NCWIT). I was very involved in the early days of NCWIT, creating the organizational foundations to allow people to be involved. As the Anita Borg Institute and NCWIT both grew significantly, I did not have as much time to participate, but I am grateful to see how the organization has grown and the impact it has on its many members.

I was honored to receive a number of awards in recognition of my contributions, including the ACM distinguished service award, an honorary membership from IEEE, named one of *Fast Company*'s Most Influential Women in Technology, as well as honorary degrees from Carnegie Mellon University and Claremont Graduate School.

13.6 What Is Next?

At the end of 2017, I left the Anita Borg Institute in the very capable hands of our next CEO—Brenda Darden Wilkerson. Shortly before leaving, I gave a talk on "What do you want to be when you grow up" [7]. It communicates my belief that a career includes multiple acts, and it is important to regularly reflect on what is most important. I continue my commitment to the technical community through board service of several non-profits. I have enjoyed learning about the semiconductor industry's maturity and the marriage between chip complexity and AI. SPAR had an AI lab that was investing in interesting work when I was there, before the AI winter. Carver's work on neural nets predated and launched many of the neural nets that are common today in AI, so my early experience had AI exposure. I am curious as to possibilities of AI in our future, and ensuring that AI Ethics are defined and managed.

I am also very interested in taking the results of the organizational focus of Anita Borg Institute, and working with companies to truly create cultures where people from all genders, cultures and backgrounds thrive [8]. I have observed that while many organizations espouse inclusion, very few of them make the hard decisions to create change. But some do, and it is important to invest my knowledge and skillset in this area to create real change. So ultimately my life's work sits at the boundary between technology and having an impact.

I created and launched a consulting practice (TelleWhitney.com), to work with organizations, and have enjoyed engaging with a number of organizations committed to change. Through the commitment of an organizations leadership and coupled with organizational change research, it is very possible to create technology cultures where all genders cultures and backgrounds can thrive. Through my board work, I am involved with creating the next generation of AI leaders (AI4All), engaging with an increasingly diverse community (Center for Minorities and People with Disabilities) and developing powerful and inclusive organizational cultures (Power and Systems).

My career is a story of entrepreneurship, from the two semiconductor startups, to the Grace Hopper Celebration, the Anita Borg Institute, and my newest work at Telle Whitney, LLC. I have found that growing an organization is rewarding in every way.

I reflect on the key lessons that I have learned along the way:

1. Take risks and go for it
2. Demonstrate resilience, and continue to show up and do the best you can
3. Understand that control is an illusion
4. Find the right mentors and sponsors
5. Realize that careers have multiple acts
6. Embrace Entrepreneurship

I am committed to learning, to exploring, and I am deeply committed to having an impact. I look forward to this next act and working with many committed leaders.

References

1. Sutherland IE, Mead CA (1977) Microelectronics and computer γγscience. Sci Am 237(3):210–228
2. Mead C, Conway L (1980) Introduction to VLSI systems. Addison-Wesley, Reading, MA
3. Whitney TE (1981) A hierarchical design rule checker. Master's thesis, California Institute of Technology
4. Whitney TE (1985) Hierarchical composition of VLSI circuits. PhD thesis, California Institute of Technology
5. Johannsen DL (1981) Silicon, compilation. PhD thesis, California Institute of Technology
6. Hedges TS, Slater KH, Clow GW, Whitney TE (1982) The siclops silicon computer. In: Proceedings of IEEE ICCC, September 1982, pp 277–280
7. Whitney TE (2017) What do you want to be when you grow up? Grace Hopper Celebration
8. Whitney T, Taylor V (2018) Increasing women and underrepresented minorities in computing: the landscape and what you can do. IEEE Comp Magaz 51(10):24–31

Chapter 14
An Exciting Journey: From Finite State Machine to Circuit Design and Ultimately to Web-Based Silicon IP Exchange

Gabriele Saucier

14.1 Introduction: My Origin

I feel like I come from nowhere.

Born in Alsace during the First World War, I was born German (Alsace under occupation) and became French after the war. My parents changed nationality four times and I changed only once. So, I definitely hate borders and wars as for one century my family dreadfully suffered from them.

My father came from a very poor family and had the reputation to be a very clever kid but who could not study because of the First World War. Once the war was over, he studied by himself and became a manager in a textile factory. He was clearly a self-made man. During the Second World War he was a resistant and died exhausted some years after the end of the conflict, leaving five children almost on their own.

Unfortunately, I remember well one of his statements "Life is tough. Girls have to work, especially as they have to earn money in order to support their brothers' engineering studies." This is what my eldest sister had to endure despite the fact that she desperately wanted to study. So, I am not sure what would have happened to me if my father would have still been there but what I did after was "Despite "in my mind.

On my side, when entering school after the end of the war I became over excited by school which was the most fascinating and friendly environment I could imagine. My feeling was that it is a place where one would really take care of children… I skipped classes, was "the best" and therefore school and learning appeared as the distraction of my life (and I did not change my mind during the remaining of my life). I decided to learn everything … even if I have to starve for it…. I decided also that I would never become a secretary or even study literature or languages despite

G. Saucier (✉)
Design and Reuse S.A., Grenoble, France
e-mail: gabriele.saucier@design-reuse.com

A. C. Parker, L. Lunardi (eds.), *Women in Microelectronics*, Women in Engineering and Science, https://doi.org/10.1007/978-3-030-46377-9_14

liking it. To survive and earn my life, science was obviously the best track…I was also passionate about understanding the world (from atom to stars). Another reason which led me to become fond of knowledge and sciences is that I hated the religious terrifying stories of my mother who believed in miracles, witches, and the devil (still common in Alsace). So scientific knowledge was obviously a protecting barrier against fear and craziness.

The next milestone of my life occurred when moving from South of Alsace to Strasbourg in a high standard girls high school. Having a hard adaptation time, my mother went to see the female dean and she was told "It is normal that she feels depressed. She is in a math/science class and girls should not be there." and I was put into an experimental science track that I hated as I preferred abstraction. This was a shocking statement that I did not absorb as it blocked me from some of the best education tracks (French "classes préparatoires aux grandes écoles") followed by my brothers.

14.2 From University to PhD

I was obviously a rebel and followed the Physics/Math course at the University of Strasbourg paid by a public grant (as an orphan) and got a Bachelor in Physics/Math very young as I was only 20 years old.

Of course I wanted to pursue my studies …I hesitated between fundamental Physics research as I was fascinated by the non-understandable world (I remember the quantum mechanics course that I followed and which was so mysterious) but at the same time surprisingly really liked learning about Computers and more precisely Boolean Algebra… A world with only two values was intriguing … especially in a top university in that field namely Grenoble. Another less defensible reason to study in this city was to follow my boyfriend, a mountain climber (who put that move as a prerequisite to any further relation …).

I joined a lab working on servomechanisms and robotics and must admit that one of the reasons leading to that choice was that I was told that I was the first woman envisioning to join this lab … So I imagined it must be a valuable track (for future real job…).

The course I preferred was Boolean algebra and sequential machines (practical version of Finite State Machine … then in robotics area) extensively used at that time in robotics applications.

The first research master project I achieved in 1962 was the programming of the Quine–McCluskey minimization algorithm stored on a ribbon with holes and executed on one of the first machines of the world, the French "Cab 500."

At that time, I followed the French academic course and consequently went through two PhDs one called a third cycle thesis and a second one, a "Public/State doctoral thesis" to be able to apply to a university professor position.

14.2.1 First PhD: Computer Based Theorem Proving

The first very good event that happened to me was to obtain as my first PhD supervisor a brilliant mathematician (Prof Jean Kuntzman) who happened to be a shy person very respectful of people including women and extremely curious about the world evolution. I reported what I learnt about asynchronous sequential machines and how to avoid hazards in state machine transition.

He explained to me that the best way to solve this problem was to take a hypercube or Boolean Lattice vision. In order to avoid hazards, for the solution, it was necessary to assign those transitions to non-intersecting sub-cubes or faces in the hypercube or Boolean lattice. I was fascinated by the cleanliness and beauty of his development as well as the access given to an abstraction level that I liked very much. Starting with a practical engineering perspective, he allowed me to get a clean innovative mathematical vision.

His second vision was that the birth of powerful computers is a major world event and we should use them extensively.

Both visions led to a "theorem proved by computer" of the universal minimal encoding of asynchronous state machines. The approach was based on an exhaustive enumeration of all possible cases on the hypercube representation of next state machine equations. It consumed an amazing computation time and was probably one of the first attempts for proving theorems by computer enumeration. The programming was done in high level programming language called Algol 68 (invented in French as was Cobol at the same time). Programming was clear, easy and just needed a good sense of reasoning. French engineering computer science school was created by Prof. Kuntzman after my PhD (So it was not yet an "engineering" discipline advised for boys of course).

The result of this work is a "3rd cycle thesis in automatisms and servo mechanisms—University of Grenoble (Institut National Polytechnique de Grenoble) on Universal Encoding of Asynchronous State Machine" that I got in 1965 with the highest grade.

It contradicted the universal Huffman statement/Theorem about the minimal number of variables required for encoding an asynchronous state machine which was worldwide considered as an absolute/universal truth by mathematicians.

The result helped me a lot as I could enter the French National Research Centre the year after and this really allowed me to go to the next step, the preparation of a state/public thesis.

It should be noted that while preparing this first PhD I had to work as a Math teacher in high schools and private schools to earn some money while having a baby son some months before the PhD. This was quite a busy tiring year. I was somewhat a scandal being a mother and working full time (did not happened before in my family).

14.2.2 *Discovering the IEEE Society*

As a researcher, the publishing duty was obvious and I liked it. I presented my result in Finite State Machine events initially in Italy and Israel, dealing with optimized state assignment and more generally next state variable dependency. I was encouraged by the scientists working on Finite State Machine (I remember Stephen Unger from Cornell University) to publish in the so-called IEEE Transactions on Computers. I never heard about it anywhere else and especially not in France. This was a discovery for me as in France nothing similar existed and it gave me a referent community, an open, cohesive, tolerant, and respectful community which laid down the ground for all my scientific and overall life!!!!

14.2.3 *Second PhD: Discovering Optimization Algorithms*

Having integrated the national research center, I could work on my state PhD to be able to apply for advanced research position leading up to a university teaching position.

I exploited the initial research path (hypercube as a Boolean lattice) to work on a heuristic for encoding any asynchronous state machine with a minimal number of internal variables. The problem consisted in embedding the state machine next equation graph in a minimal sized hypercube.

The representation of the hypercube, sub-cube, faces, and face intersection within the hypercube was very well mastered. The last part of this PhD, inspired by the hypercube representation, led to an efficient algorithm for detecting isomorphism between graphs. Mathematicians proved later that it was the best heuristic known at that time. As a fun fact, I remember the surprise of a Russian mathematician discovering that I was a (young) woman. A real shock for him!

This led me to the so called Public/State thesis on "Encoding of asynchronous machine."

It should be noted that this work was done while a second baby, a girl, was added to the family in 1967. My husband went to his military service promptly after that birth. The 1968 year was agitated in France. I joined my husband in Alger and was teaching at the institute Polytechnique in Alger to earn money.

In 1970, I came back with a third baby (a boy) and presented my state thesis in Grenoble.

14.3 Heading a Research Team

Some months after my state thesis, an unexpected event happened. A university professor position was open and dedicated to what professor Kuntzman called "Mathematics for hardware." The teaching scope was Computer architecture, Boolean Algebra, and logic design. I was nominated due to my IEEE publishing and to the fact that there were few scientists in this area in France. Moreover, there was an internal tension among other candidates and therefore, I got that permanent professor position at the University of Grenoble (INPG). I became the youngest professor nominated in Computer sciences as I was elected at 29 years old.

Prof. Kuntzman at that time created the first engineering school in Grenoble dedicated to computer sciences. He asked me to take over the Boolean Algebra course and discovered what is an amphitheatre course… and it was OK for me. Preparing for his retirement, he asked me to take over a research contract on test generation of Electronic systems and definitively pushed me to concentrate on hardware-oriented research. He was fond of relations between industry and university research (unusual for a mathematician) and in no way he pushed me to more fundamental research.

My teaching was extended to Logic Design and Computer architecture.

14.3.1 Test and Fault Tolerance Period 1970–1980

I joined the IEEE fault tolerance working group covering reliability and test generation and remember a very friendly stimulating international community. A strong focus was given to functional testing (online test and self-checking controller up to microprocessor testing) and raised a significant number of PhDs in a growing lab. The publications in this new area together with the PhD results triggered my promotion to the IEEE fellow grade under the sponsorship of Prof. Ed McCluskey from Stanford University.

The positive aspect was that the research in test and fault tolerance (including hard radiation) led to a long research track.

14.3.2 Microelectronics Invasion and EDA Research 1980–1994

A double influence oriented at that time research lab topics:

- Locally STMicroelectronics was created and became a worldwide leader in semiconductors. Grenoble could seriously believe to become a French Silicon Valley.

- At an international level, Carven Mead brought an abstraction level to electronic design (symbolic layout) and we could suddenly all become design engineers starting from a logic level. This seemed just fabulous and all EDA and electronic design community and engineers remember that transition.

Combining Boolean Algebra and finite state machine knowledge with early experience in heuristics and adding this new abstraction level led to an explosion of research and results. The results were *de facto* oriented on the production of the most efficient EDA (Electronic Design Automation) tools.

In parallel the demand for teaching in Boolean Algebra, Logic design, and Computer architecture became very high and my courses were extended from Computer Sciences students to programmers, and microelectronic designers. Basic notions of circuit and computer architecture were of interest for a broad spectrum of students.

Overall great progress was achieved by adopting an integrated approach taking into account the final topological view during the Logic optimization phase. Optimization at successive design steps without coherency just does not work as we all know. A famous paradigm I defended was "Do not minimize the number of literals in logic expression but look at your global optimization process."

14.3.3 Preliminary Result

Preliminary first research at the topological level led to provide an optimized floor planning approach (channel routing algorithm got an early best paper at DAC Design Automation Conference).

A second result was the proposal of a mapping algorithm on CMOS gates inspired by my PHD work on Graph isomorphism… The proposed algorithm started with a Boolean expression (represented as a tree with optimized input order), applying a covering method identifying first the "most complex" gates pattern and then other library elements.

Following some more achievements at the topological level, the most innovative results dealt with logic minimization predicting and optimizing the routing area.

14.3.4 Minimizing the Global Circuit Area

A very early paper on optimized synthesis of complex CMOS gates attracted my attention to the importance of ordered inputs (due to metal strips crossing those complex gates).

Due to the collaboration with STMicroelectronics and access to actual layout the importance of interconnect area with respect to the active logic area was obvious very early. Optimization of new Boolean expressions (called lexicographic Boolean expressions) based on the ("best") ordering of inputs led to the best global circuit area including the routing area which appeared to be the major area consuming part…

A surprising side result of that approach was that the discovery of the best ordering of input variable targeting routing minimization was also the best for minimizing the complexity of a Binary Decision Diagram (BDD) representing a Boolean expression. A wide range of applications were triggered by the progress on BDD.

14.3.5 *Top Down Design Integrating a Vision of the Final Circuit*

Two other experiences consolidated the need for a global top-down view especially when targeting pre-structured circuits (programmable circuits). Both the starting Boolean expression and the mapping algorithm just stick at the topological structure.

- The first research work concerned synthesis on FPGAs and mainly the Actel FPGA (Field Programmable gate array) devices where the final structure was very well adapted to binary decision diagram. Thus, the synthesis method just started from such a diagram which was covered very easily with the Actel physical cells. And overall the open mind of US companies with small French research group was one of the most fascinating aspects and the collaboration one of my greatest experiences. Similarly, decomposition and mapping algorithms for Xilinx cells proved the soundness of the approach.
- A similar experience targeted another class of programmable circuits, the CPLD, structured as a cluster of product terms in the circuit. Thus, a blind preliminary optimization has to be replaced by a minimization step identifying exactly those clusters of products terms depending on the same subset of inputs of course. Excellent results were obtained

14.3.6 *And Final Achievement: The "Best" State Assignment of Finite State Machine*

Of course, going back to state assignment after 20 years, it was more than tempting to revisit the state assignment issue. If the purpose of my PhD was to find the minimal number of internal variables to encode an (asynchronous) state machine, the goal here was to minimize the next state logic. The hypercube vision was still the solution: If you assign the next states of a current state of your state machine to a sub-cube of a hypercube (of minimal dimension) the next state equations will be simplified.

Applying this heuristic all over the state diagram leads to an excellent result. Patented with the help of the company "VLSI Technology Inc.," the result was also transferred to Electronic Serge Dassault. Those transfers were received as a big reward… The pleasure to see it real… and over all the big competitors. Berkeley University (Professor Brayton) acknowledged that this was the best result over all the famous benchmark cases official at that time.

14.4 More Experiences and More Connections

14.4.1 European Research Projects and University/Industry Relation

European Grants through cooperative research projects were the second source of support of all our research activities, IEEE support that I already mentioned being the first one.

A first project (Euripides) mainly headed by Germany targeted the creation of European EDA tools. It never happened but brought a significant amount of money to pursue our research.

The next European contracts we joined were conducted by industry and brought very unique design experiences. We were invited to join advanced industry design teams due to our experience in synthesis and circuit architecture.

An amazing project was the so-called wafer scale integration project initiated by Bull and then taken over by STMicroelectronics. As its name suggested, the goal was to design a system on a full wafer; a major obstacle was the manufacturing yield.

Within this project we fully designed and tested as test chips:

- A full wafer high yield memory
- A defect tolerant RISC processor (already)

In a next phase we participated in the design and implementation of a full wafer of:

- A network of neural processors (collective patent)
- A systolic array of image processing

These experiences were fascinating and consolidated a practical industrial view.

Nevertheless, the demand of those industry collaborations was very demanding. It brought a lot of money to the university but the researchers needed to be reliable as actual project participants and this led to a hot and painful issue.

Initially the university president strongly supported the participation to those projects (bringing in a lot of money). A new president (pure mathematician) was more sceptical. Some researchers (especially a woman) complained about the pressure of applied research and ask for my resignation. Due to industry pressure, a compromise was found by splitting my lab. This is a good example of the difficulties met in university/industry cooperation.

14.4.2 International Connections

Of course, universities are ideal environments for establishing international relations. I mentioned already the importance of the IEEE society. I was especially fascinated by the invitations and relations with Russian academicians (Novosibirsk).

I was part of an official French mission in 1977 to establish relations with the newly opened China and was invited to teach VLSI design to 150 university professors in 1979 in ChengDu.

Having at the same time a contract with the US army (Dayton USA), for exploring the importance of Petri nets), I got an inspection from the French government to verify that I was not an international spy... I was not ... I was just a naïve university teacher happy to get grants or invitations.

14.4.3 Events

Participation in conferences was a normal activity for any university professor and I counted more than 300 publications. Nevertheless, over the years the creation of our own workshop was tempting to be sure to focus on precise topics that were evolving over the years.

I created the EuroAsic conference about ASIC design which was a new topic. It became quite successful and when it looked suitable, I promoted the merge with EDTC and EDAC conferences.

After that I created a non-interrupted series of events starting from synthesis up to IP-based design and today the well-known IP SoC events rotating worldwide (USA, Europe, China).

14.5 Entering the Business World

14.5.1 1990–1997 Creation of IST (for Innovative Synthesis Technology)

The big surprise was that synthesis became a major business in a very short term and a very aggressive player (namely “Synopsys”) appeared as the leader in the landscape. Due to the long-term background at Synthesis research (25 years at that time) and accumulated innovation I asked myself the following question: “What should or can we do from France in our environment?”

In parallel, locally, the push for technology transfers to industry became preeminent.

As I mentioned above, the demand from industries for a transfer of our “jewel” asset (namely the state assignment algorithm) to demanding companies pushed the creation of a first start up. Created as a "researcher community" a better action was needed swiftly. The confusion between university and the business world was too hectic.

Thus, a more structured attempt incubated by the university led to the creation of company IST for "Innovative Synthesis Technology." The first customer support came from Japanese companies envisioning an alternative synthesis tool to the monopoly situation of Synopsys and they were extremely supportive. It seems obvious that there was a chance for an alternative synthesis company which made sense for the IC market at that time.

Later on, focusing on synthesis on FPGA became a choice for two reasons:

- We demonstrated excellent results
- Possibly it was better to avoid "head to head" competition with the growing Synopsys.

By the way this choice was certainly not the best" business" decision. Note that prototyping methodology on FPGA was an attractive "side track" and led to a straightforward approach immediately used by companies like STMicroelectronics to prototype their micro controller.

Upon arrival of a French investor, a very bad decision was made, namely the import of a US CEO absolutely incompatible with a French start-up company, fully unable to write any business plan and who promptly spent all the money. I was kindly invited to withdraw from any business decision from the first day of his nomination.... This led promptly to a "Chap. 11" status (painful experience) and I was left with neither technology, nor people.

In order to "survive" or recover, I reacted vigorously. I certainly did not want to retire and therefore created some months later, my next company called Design and Reuse.

14.5.2 Creation of Design and Reuse (1997-Up to Now)

At that time, the paradigms of "design-reuse" and "(silicon) IP" became popular in the EDA conferences and intrigued me.

Several facets triggered this initiative:

- A reusable design block (or IP) encapsulated all the knowledge on design flows and EDA as you deliver all views (RTL/synthesis scripts/netlist/layout) of your block. It was like a summary or a global vision of such a block based on all my knowledge in the field over the 30 previous years.
- Reuse or IP Exchange was associated with web-based collaborative design. The exploding interest in web information exchange was very exciting and obviously in my whole life everything new obviously fascinated me. In addition, it helped me to recover from the loss of my "asset" (tool and people) in synthesis and EDA.

- IP licensing became obviously a worldwide business and the designers, all over the world, were not yet so familiar with web techniques. Thus, there was a lot to explore or to promote.

Therefore, a next start up was created in 1997 namely Design and Reuse with the following "business "decisions:

- It appeared nearly impossible to launch a company from France due low-level start up support at that time.
- We should rely on our own web support for communicating and marketing whatever our products were (no investment should be required).
- In other words, we should be autonomous (suspicious about VC after the last experience).
- To clarify the situation, I definitely retired from my university career that I ended with the maximal grade (exceptional Class Professor) which was a big chance.

We immediately restarted with the following actions:

- Identifying worldwide IP offers and creating an IP list with Philippe Coeurdevey as a student.
- The goal was still unclear. Should we design our own IP on FPGA? In order to stay "motivated or alive…," I borrowed a booth wall from another company at DaC (Aptix) to show our first list of IP.
- Posting that list of IP on a web portal (fascinating at that time) www.design-reuse.com was the next step. This portal was conceived as a marketplace where providers and consumers could meet.
- At the same time and in parallel elaborating modifiable and exchangeable IP Catalog technology (IPMSTM) based on Java and XML (leaving definitively previous programming technology in C) makes me feel comfortable as a technology asset (Not only scripts).

The starting and ramp up was unbelievably fast and successful:

- The web site quickly became worldwide famous and attracted sponsorship.
- I should be noted that when opening that website I received a threatening phone call from the USA telling me that I should not open such a website if I did not want to get into trouble!!!
- However, we reached promptly close to 1M of visitors (Google statistics).
- EE Times (CMP at that time) invested and acted in fact as a launching VC.
- Industry contracts to hold and update, over the web, external IP Catalogs in a large enterprise were signed the first year (Alcatel being the first)
- In 2000, D&R got the "New Technology Trophy" from the French well-known journal "l'Express". It was a highly recognized award for companies entering aggressively the new web world.
- Well recognized VCs contacted and met us but we were unable to envision or provide a web-based business plan as web user entry was free and the IP/Asic is still a niche at a business point of view.

- Acquisition offers were made but as I just wanted to be busy during my retirement, I did not want to give up even for money.

In 2009, I was nominated as part of the ten Top Women in microelectronics by EEtimes and got the silver recognition from IFIP.

14.5.3 And Finally IPMSTM (IP Management Systems)

The creation of a website was just "not enough for me" and I went on in pushing the IP management technology for creating a collaborative design environment inside enterprise intranet.

After the first application—namely creating and exchanging catalog data—the focus was on creating the so-called IP Reuse Station aimed at putting in place an Enterprise Infrastructure allowing designers to easily reuse what has been designed ahead by others.

The configurable platform was based on the idea that accessing a database directly is cumbersome and that a layer namely a "XML format" on top will allow to look at the Data through different (meta data) views and facilitate actions on those data (access to actual design data, download or exchange of actual files).

Fast recognition came from major players and the installation and support of a Reuse Station at Cisco Inc. in 2000 allowing key IP exchange among their remoted design centers was a great achievement within a friendly cooperation.

This platform was so configurable that applications were just infinite. The most amazing being probably the royalty IP management for a major (GAFAM) player and Software license management integrating user monitoring for an Israeli military enterprise.

14.5.4 And Where to Go?

It may appear that the business vision around the last D&R "innovation" namely IPMSTM leads to a deadlock for a small French company

Solving advanced management issues for major players was oblivious very satisfactory.

But… Selling intranet management platforms can only be proposed by extra-large IT or service companies and is de facto in perpetual competition with internal corporate IT.

IPMSTM shows that innovation power of extra small entitles is infinite… but the passage to industry IT products is a different story

Thus, D&R using its own asset to manage efficiently its three synchronized web sites is the last challenge? As well as supporting the worldwide IP community with big demand from China is already a huge challenge and maybe the last or not.

14.6 Annex: "Professional Me Too" in Microelectronics?

I will certainly not comment on too personal incidents in my professional life and associated travels. Due to my origin and my "survivor" profile, my overall vision was:

- what happened to me must be my responsibility or worse my fault....
- Looking at the worldwide woman status, I can only be happy with the fact that I could study and have a professional life and I should not complain.

Nevertheless, listening to younger women, I may relate some "professional" facts that could only happen because I am a woman. So, let me illustrate briefly two stories related to two professional periods:

1. Contract and university competition
2. Experiencing Business Competition

14.7 Story 1: A Research Lab Industry Contract and University Competition

While heading my research lab, some results (especially test generation and later on optimized state assignment for sequential machines) were known and appreciated by industry. Thus, the research lab could get industry research money on those topics.

A contract was signed with ESD (Electronic Serge Dassault), at that time, about test generation. A high-level executive, responsible for following up the contract called me one day and told me that he had to report about a very annoying fact. His wife (mother of 4 or 5 children if I remember well) went to a deep depression before telling him the reason. She was harassed by phone calls from an anonymous person telling her that in reality the contract that her husband was conducting with Grenoble University was just an excuse to hide an affair between this executive and myself. This executive was an extremely polite well behaving person, very respectful of research and persons whom I met exclusively for working meetings. Dassault investigated the phone origin and told me that the call was originated from our research center consequently from a competitive lab.

The executive told me that Dassault management was shocked and offer to help me for legal pursuit that they strongly recommended as I could be in danger if somebody behaves like that in my close environment. Equally appalled by the news, I went to the highest-level research lab manager and told him the story. The answer was "I am absolutely not interested in your love or sex life" and that was the end at the university side. Dassault was just horrified and stated that university competition can be wilder than in industry and obviously has no limit.

Besides I had to experience the difficulties of heading a high-tech research lab. I saw a researcher of the lab getting the best paper award at DaC and just taking out my name as his PhD/lab manager. And many other similar incidents.

14.8 Story 2: Experiencing Business Competition

I obviously became nervous to see that synthesis technology the passion of my life became one of the most powerful business enablers at that time in the semiconductor world. Frustrated, I spent my energy to compete with my weapons, concentrating on the famous benchmarking of synthesis tools focusing naively at that time at minimizing the area of logic in IC. In the environment of Grenoble I had access to actual silicon layout and could demonstrate easily that the point was not to decrease "the number of literals in Boolean equations" very fashionable at that time but to have a top-down view and as an example for sequential logic synthesis mastering the state assignment to predict Boolean expression minimization and for logic expression to predicting the routing. Results were just amazing and I encountered a surprising incident:

- After a session at a synthesis workshop where I displayed the result, a well-known company research member grasped me like a parcel and threw me in a swimming pool with my bag (passport and phone in it) stating "I just can't anymore watch your benchmark results"…. It looked like a huge scandal and I just decided to laugh at this incident. After all, as I said, I like water and can swim… I was blamed by the US women community for my lack of reaction and after years, I think that they were right.
- I asked the CEO about the replacement of my phone but was told that he was not aware of such an incident. I did not measure at all the legal implication of such an incident could have had in the USA….
- Coming back to France, a French representative of that company came to me in the airport and told me that if did not stop showing such type of research results, I could encounter major problems!!!!
- I addressed a question to that CEO in an official meeting where he claimed the importance of benchmarking and triggered a very violent reaction in front of journalists.

Dedication: To my granddaughter Julia who convinced me to write this chapter.

Chapter 15
The Quest for Energy-Aware Computing: Confessions of an Accidental Engineer

Diana Marculescu

15.1 Unlikely Beginnings

I grew up in an Eastern bloc country—Romania—that in the 70s or 80s was the opposite of a champion for democracy. I quickly learned that the only way up or out was through education; that was what had brought my parents together in the country capital, my mother from a nondescript village and my father from a small provincial town. Neither of them had gone to college, so my older sister and I ended up being first in our family to get a college degree and later, to go to a graduate school and get a PhD.

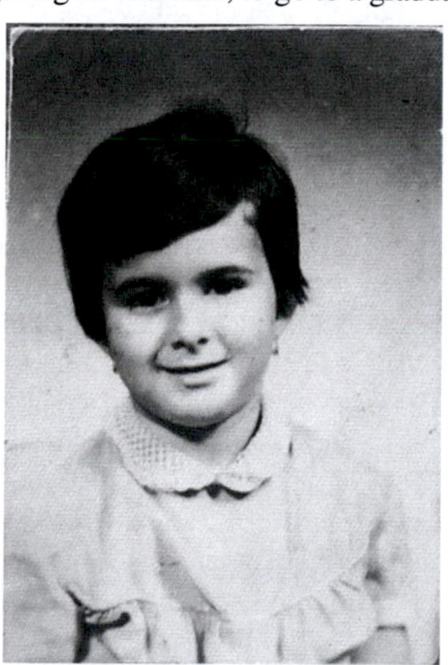

Diana at about 5 years old

D. Marculescu (✉)
The University of Texas at Austin, Austin, TX, USA
e-mail: dianam@utexas.edu

A. C. Parker, L. Lunardi (eds.), *Women in Microelectronics*, Women in Engineering and Science, https://doi.org/10.1007/978-3-030-46377-9_15

Unlike many aspiring engineers who grew up surrounded by relatives working in the field, one of my few connections to tinkering with things was helping my father put together our Christmas light strings every year: handcrafted from scratch and individually tested, since no store was selling them pre-made in Romania at those times. As a young child, I was shy (and painfully so) to the point of making those who did not know me think that I could not speak. I could (obviously!) speak and very eloquently so, as my school years eventually showed. As a young elementary student, I loved math and brain teasers that my mother was providing in copious amounts. My grade school was an unremarkable neighborhood establishment where, for some unknown reason, a few teacher gems had ended up and I had the luck of studying with them. My math teacher, Ms. Veronica Marinescu, is one of the main reasons I ended up on an engineering career path. She taught math with rigor and tough love, yet she had a league of student followers who kept in touch with her after they had left for high school or college. I later realized many of them were women in STEM careers (like me), subconsciously inspired to take an unlikely path forward. My Romanian literature teacher, Mr. Ioan Zaharia, was droll and foul-mouthed—at least by grade school standards—but he instilled a love for reading and writing in his students, and I turned out to be an eager audience. My English teacher, Ms. Nadia Kotovsky, shared her love of teaching a language that was and is universal, yet at odds with what the Romanian politics were at the time. Living in a police state was not easy, but I enjoyed surprising liberties enabled by access to good education that allowed me to imagine what lay beyond the grim reality we lived in. I dreamed of changing the world and how we interacted with each other. I remember how in early elementary school years I longed for more time with my mother who was working long shifts as a physician assistant, so I boasted that as a grown-up, I would built robots to do the work for me so I could spend every day with my children. Ironically, I did not end up building those robots (nor spent every hour of every day with my children), although my research has brought me closer to the field of machine learning (ML) and artificial intelligence (AI) that is shaping today's robotics applications. Later as an adult, I took my love for math and science to computers and engineering, while never leaving behind my fondness for literature or arts. To this day, I cherish the time spent reading, listening to music, or walking through a gallery, but it is easy to see what I ended up choosing as a path, helped by gifted teachers who transcended their fate in a closed system, by educating and inspiring others.

My high school years were less remarkable, perhaps also because of the bleakness of those times. In mid 80s, the Romanian society had reached a turning point, marred by deep discontent and muted social unrest. Food shortage and scheduled power outages made for a teenage life that was light-years away from what most high schoolers are used to today. Like in other Eastern bloc countries, those who were vocally protesting or unsuccessfully attempted to leave the country were publicly suppressed or imprisoned. In silent protest, many people, including teenagers like me, took refuge in reading the banned "1984" and "Animal Farm" by George Orwell or Ray Bradbury's "Fahrenheit 451." I was in a high school with a science-focused instruction, but those times had forced many of us to think holistically beyond math and science and embrace humanistic endeavors for potential society change. My high-school years have also made me understand how one can learn and move from failure. During the comprehensive exam at the end of tenth grade (largely a "weeding out" exercise that used to push bottom students to other concentrations), I had received an uncharacteristically low score in my Physics exam, ranking high enough to stay in the highly sought after concentration, but a far cry from ranking first in the high school entrance exam two years back. To make things worse, my best friend at the time was ranked first! While we stayed friends through the first few years of college (and she also became an engineer), it was a sobering experience to cope with what I perceived a failure and a loss to someone I considered very close. Over my years as an academic, I learned that recovering from setbacks and being graceful when your friends or acquaintances' win are your losses (typically the case in a zero-sum scenario) are some of the most important skills one can learn and use.

15.2 From Math to Computer Science

For much of my time in middle and high school, I thought I would become a mathematician—the only thing I needed was paper, pencil, and a problem to solve. I vividly remember solving Euler's nine-point circle problem late at night when an idea struck. I was not alone in this—Ms. Marinescu had a summer math assignment set for those eager to take it and some of us were fiercely competing for the position of the top problem solver. That spilled into the weekend Math Circles some of us attended throughout middle and high school where we would prepare for the annual Math Olympiad. I could not imagine not doing math, so that seemed like the right path for me, but I turned to computer science which was growing rather fast as a discipline at the time.

I was admitted through a demanding national entrance exam at the top engineering school in Romania (now called "Politehnica" University of Bucharest). I was third, preceded by two other women and followed by only about 150 other candidates who were able to score 86% or higher. In that place and time, girls thought computer science, electronics, and engineering were just as good majors as any, and many—though not all—thought girls could do it as well as anyone else. Much of the scoring and grades throughout college was public, and many instructors made a point to praise those who were at the top. One of our advanced linear algebra professors had the habit of calling students in his classroom in decreasing order of their exam grades. In one of his exams, I was the first to be invited in, followed by a few of my male colleagues. "Ladies first is good if you want to show you're well mannered, but a math exam isn't such a place," he chuckled. I realized years later that I may have disrupted a long tradition of having men enter his classroom first to get their final exam top grades—and I could not have been prouder of that.

Although I was a computer science major, I was a first-generation college student and had not used or seen a computer except in movies or magazines. This is unheard of nowadays, but it was par for the course in the Romania of the 80s. During my first three years of college, I went from programming using punch cards in Fortran and Pascal, to programming in C on a RSX-11M shared-time machine and eventually on a PC running MS-DOS. That may sound arcane (and it may have very well been), but it instilled good habits in the untrained programmer I was at the time. Indeed, one had to compile their code in their head before attempting to submit the punch cards or use the allotted time in a shared-time system. Our professors were considered local geniuses given their prowess in both theory and practice of computer science, but were largely unknown to and isolated from the outside world where computing was thriving. Access to non-Eastern bloc publications was limited and one had to go through multiple approvals before they could get access to and read IEEE Transactions papers or ACM publications. Publishing in those venues was possible at times, but traveling for a presentation was as rare as a unicorn as it typically ended with researchers defecting to the Western world. However, totalitarian regimes in Eastern bloc countries were approaching their end. By the time I graduated, all physical and virtual political walls had been dismantled and I was determined to not only go up but also out, out of a place frozen in time, whose people were paralyzed by half of century of inaction.

15.3 From Computer Science to Computer Engineering

Diana and Radu in New York City in 1993, days after landing at JFK.

After graduating with a computer science degree in summer of 1991 and spending the rest of the year as a software engineer, I felt my place was not in the software development world. I had kept in touch with my college professors, so when a teaching assistantship opened up, I went for it in January 1992. I was assisting with several courses, including one on Formal Languages and Translators taught by my former undergraduate thesis advisor, Prof. Irina Athanasiu. At the time, the department was a hotbed for young computer scientists and electronics engineers who were teaching or assisting with courses, yet aspiring to become PhD candidates and do high caliber research somewhere in US. European Union was not yet as hospitable to Eastern European researchers as it is now, so most of us were busy taking GREs, TOEFL, and preparing for the long haul road to a US graduate degree. My life was about to change as I not only went on crossing the Atlantic embarking on a journey for a Ph.D. degree but did that with my better half, my future husband, and my first collaborator in research. Radu was already a budding researcher who had defied the odds of being in an Eastern bloc country by publishing his work outside the borders—yet not being able to travel and present it. A fateful link with Irina Athanasiu had connected us for what has ended up being more than a quarter century together. During the fall of 1992 and much of 1993, we have shared the ups and downs of applying for grad school from a country that had been isolated for decades from the rest of the world. This was before the era of electronic applications, so we were at the mercy of postal services to get our applications sent on time. Some schools had sent their application materials using ground service (unclear what that meant when there is an ocean in between) and many of them reached us after most deadlines had passed. Of the handful, school applications we were able to submit and be admitted for, none had both of us together or even in the same time zone. I took the admission offer from New York University (NYU) at the Courant Institute for Mathematical Sciences, and Radu went on to join Prof. Mel Breuer's team at University of Southern California (USC) in Los Angeles. We traveled on one-way tickets to New York City's JFK airport, largely because we were short on cash and only carrying our suitcases and less than $2000 borrowed from relatives and friends. But the one-way trip marked the beginning of our lives in the country we have called home for more than 25 years.

At NYU, I had the privilege to be a teaching assistant for Prof. Alan Gottlieb's class on Programming Languages and take Prof. Richard Cole's Advanced Algorithms class. I was not doing research yet, but I truly enjoyed the algorithmic world of computer science where one tries to find best ways to solve a problem without a human in the loop. I thought I could apply my newfound enthusiasm for algorithms and optimization to solving practical problems. I applied for PhD in natural language processing (NLP) at USC and other Southern California schools for Spring 1994, but funding was hard to come by. I still decided to move West and join USC, and in a twist of fate, both Radu and I ended up with our future PhD advisor, Prof. Massoud Pedram, who was working in a new area of Electronic Design Automation (EDA) that dealt with automatic tools for low power electronics and design. It had the right mix of algorithms, optimization, and practical applications, so I switched to getting my PhD in computer engineering and working in Computer-Aided Design (CAD) for low power systems.

Diana, Radu, and Andrei (middle) in 1998, after the Ph.D. hooding ceremony at USC.

The focus of my research during those years has been on developing efficient tools for modeling, estimating, and optimizing power consumption of electronic systems. During the early 1990s, it had become obvious that power dissipation was the leading limiter on the design of future integrated circuits and especially microprocessors. Moore's Law scaling had started to run into an increasingly difficult set of power and energy challenges as semiconductors continued to (try to) shrink. These were the golden years of CAD when most foundational work happened and helped establish EDA as a first class industrial partner for the semiconductor industry. During my doctoral studies at USC (1994–1998), my work on information theoretic measures for power modeling in digital systems and on probabilistic analysis of finite state machines and stochastic sequential machine synthesis enabled orders of magnitude speed-up in power estimation of digital systems. As one of the architects of the USC-based project POSE (Power Optimization and Synthesis

Environment), I had the rewarding experience of working on real examples and test cases together with engineers from Epic Design Technologies (subsequently acquired by Synopsys) and Toshiba Corp., thus proving the practical suitability of my work in a real setting.

I defended my thesis in August 1998 with Profs. Massoud Pedram, Mel Breuer,[1] and Michael Arbib on my committee and decided to join University of Maryland, College Park as an Assistant Professor that fall. By now, my husband and I had a 2-year-old son (Andrei), so embarking on our academic careers while having a three-hour flight between us (Radu joined as an Assistant Professor at University of Minnesota) posed again a challenge, which we hoped we would eventually overcome.

15.4 Starting the Quest for Energy-Aware Computing

After spending one year and a half at University of Maryland as an Assistant Professor, I (as well as my husband) was presented with an opportunity to join Carnegie Mellon University (CMU) in early 2000. Although I had interviewed and was offered an Assistant Professor position at University of Minnesota in December 1999, the possibility of both of us joining CMU was alluring. CMU was one of two institutions where EDA had been supported by the Semiconductor Research Corporation through a Center of Excellence in EDA. That, coupled with the school's overall reputation, made us take the challenge and move. We were now part of a two-body "solution", not "problem," as the CMU ECE Department Head (Prof. Pradeep Khosla) who hired us used to say. I had been now on the academic job market twice, once during a long and grueling process in 1998, and then later in 2000 when we have only approached institutions that seemed open to hiring both of us. While this may seem like a natural way to approach dual academic career situations today, twenty years ago many departments were afraid to make the step. While interviewing in 1998, I had experienced perhaps all stereotypes about women in STEM known to men, from "Who has done what?" if my papers had multiple co-authors, to "Are you serious about this job?" and "How will you solve your two-body problem?". That was largely a surprise as I did not expect that social norms and perceptions of the late 90s would interfere with my ability to secure an academic position. In fact, I was mostly oblivious to gender stereotypes or the lack of women in my discipline as even in my years at USC I had women in my classes (perhaps because of my curricular choices—like Prof. Seymour Ginsburg's class on Formal Languages) or in our research group (there was another couple and several other women in Massoud's research group of about 15 people). My 2000 academic

[1] On an amusing note, Mel did not agree to serve on my thesis committee until after I had presented the topic and pretty much "defended" it for him during a lengthy meeting. It goes without saying that many of the ideas that came up during that meeting made for a better thesis write-up.

job market experience was different, mostly because I decidedly did not apply to institutions that would not be supportive of dual academic careers.

Looking back, twenty years ago, I was the first hire of the then newly appointed ECE Dept. head. I was the only woman on tenure-track, and twenty years later, I am the first woman to have been hired and have gone through the ranks from assistant to associate and full professor. The department has grown from the sole woman faculty on tenure-track to eight (or almost 20%) in recent past, and I am proud to have been part of this period of growth and renewal.

At CMU, my work has expanded from design tools to design itself, and from dealing only with power consumption constraints, to expanding to emerging platforms and application as well as including design reliability and nanoscale semiconductor variability as additional constraints. During the first decade of my time at CMU, I had the privilege of working with outstanding students on pushing the envelope in energy-aware computing.

15.4.1 Energy-Aware Computing and Low-Power CAD Software Tools

During the early 2000s, computer architects had started to realize that the quest for increased performance had to be coupled with addressing the increasing impact of power consumption. At the time, I was among a handful of researchers that were building both software chip design tools and novel computer architectures to address these challenges. My work in this space was among the first to address the need to provide mechanisms for energy-aware computing at the microarchitecture and architecture levels. The main idea was to exploit variations in hardware resource usage within software applications to allow: (1) dynamic reconfiguration or selective shutdown for energy reduction and (2) optimal caching and reuse of repetitive, pre-scheduled instruction sequences for further power savings. My work in this area had demonstrated up to 40–60% reduction in energy-delay product and is still cited in the area of power management at architectural level.

A few years later, my group and I have explored how energy can be improved via fine-grained power management. Power is proportional to both operating voltage and clock frequency; to minimize power, we would prefer lower voltages and slower clocks. But these are anathema in high-end processor design; we cannot afford the enormous performance loss. A better solution is to adapt voltage and frequency on select subsets of the architecture, independently tuning each voltage–frequency island to the best performance/power point. My research in this direction was started following early theoretical work proposing a Globally-Asynchronous, Locally

Synchronous (GALS) design style for complex single-core architectures; GALS theory proposed that each island runs at its own speed and voltage. In an early and now widely cited paper at the 2002 ACM Int'l Symposium on Computer Architecture, we have shown exactly how GALS theory and voltage–frequency islands could be used on a real processor core, and with what precise impacts on both power and performance. Together with my students, I have subsequently extended these ideas to System-on-Chip (SoC) designs that integrate multiple hardware elements in a plug-and-play fashion and provided practical approaches for mixed-island interface design for robust on-chip communication. My work has inspired projects on dynamic power management at other universities, and a GALS architectural simulator developed by my group has been downloaded and used by dozens of academic or industrial research teams. Our results have been encouraging, as we demonstrated up to 5–10× power reduction in flagship computer architecture and low-power VLSI system design venues.

Although our work has won several best paper awards and nominations (Asia South Pacific Design Automation Conference in 2005, ACM/IEEE Design Automation Conference in 2007 and 2008) and is still an important reference for power management (e.g., our ACM/IEEE Intl. Symposium on Low Power Electronics and Design of 2009 power management work), it was not without challenges. SRC and some industrial partners had funded our early work, but National Science Foundation dismissed it since, as one reviewer put it, computing systems at the time relied on single clock processor which were working just fine. That actually stopped being the case in the 2010s when multi-core systems with multiple voltage–frequency islands became prevalent.

15.4.2 Ambient Intelligent Systems and Electronic Textiles

Not everything we would like to design is based on silicon or on conventional processors. Furthermore, emerging devices and unconventional platforms will likely exhibit inherently larger failure rates than mature silicon technologies. My research in the first decade of 2000 has also addressed the need to properly characterize these from the joint power-fault-tolerance perspective. Target applications have included e-textiles, ("wearable" fabrics with local monitoring and computation, as well as wireless communication capabilities) and ambient intelligent systems that are able to self-manage in the presence of changing operating conditions. My group developed the first design flow for e-textile applications and laid out the foundations for self-management mechanisms for ambient intelligent systems that could revolutionize the way we interact with our (otherwise, "non-smart") environments. I also introduced the concept of dynamic power/fault-tolerance management (DFTM), which extends classic dynamic power management by integrating fault-tolerance based on reconfiguration. This work has had remarkable visibility: our e-textiles work was featured in a cover story for IEEE Spectrum (2003); I was interviewed or consulted for articles in Newsweek International (2003), Discovery Channel News Online (2006), and IEEE Spectrum (2007). Having your story on the cover of the main IEEE magazine lends itself to achieving some sort of popular status that regular scholarly work rarely does; yet our work was also deeply technical and in some sense, esoteric and too early for those times. It will be more than 15 years later that the Met gala would feature true e-textiles worn by celebrities and serious efforts would be put in setting aside federal funding in this field.

15.4.3 Reliability- and Variability-Aware Computing and CAD Tools

As Moore's Law has driven semiconductors toward atomic dimensions, it has become impossible to regard basic hardware components as deterministic. Wide ranges of statistical variability are now the norm. For example, chip design flows relying on deterministic timing models are no longer appropriate, thus requiring a swift transition toward a probabilistic design. My research was among the first to address this issue at microarchitecture and system level, by developing models for performance and energy variability that can be used by computer architects and SoC designers. This work on energy and variability interactions has already suggested that up to 12% of performance can be reclaimed by correctly managing design variability, and for throughput-aware systems, chip yield may be improved by more than 80%. In the area of reliability-aware design and tools, my group has developed a framework for probabilistic analysis of soft-error rates in combinational and sequential circuits that has allowed for exact evaluation of all error masking effects (logical, electrical, and timing) with less than 5% error and 10,000× speedup over conventional circuit simulation. My group was rewarded at the time with two Best Paper Award Nominations (2005 Design Automation and Test Europe and the 2007 IEEE Int'l Symposium on Quality Electronic Design conferences).

15.5 Recent Endeavors in Energy-Aware Computing and Beyond

Diana with her Ph.D. students in 2009 (l-r): Da-Cheng Juan (now Research Scientist at Google), Kai-Chiang (Alex) Wu (now Assistant Professor at National Chiao-Tong University), Siddharth Garg (now Assistant Professor at New York University), Diana, Sebastian Herbert (now Principal at DC Energy), and Natasa Miskov-Zivanov (now Assistant Professor at University of Pittsburgh).

My research in the early 2000s had dealt with analysis, modeling and design of computing systems in an energy-aware fashion by adapting their operation to the applications they run, the users they have, or the environment they operate in. My most important contributions to this day are from that decade, attested by recognition by peers, and support from a variety of federal (NSF, DARPA) and industrial sources (SRC, Intel Corp., Cisco Systems, Samsung Corp.). As a culmination of my work in the field of energy-efficient computing, I was elected an IEEE Fellow in 2015, an ACM Distinguished Scientist in 2011, an ACM Fellow in 2019, and on the IT Honor Roll by IT History Society in 2012. As a further recognition of my work, I was selected as a recipient of the Marie R. Pistilli Women in EDA Achievement Award in 2014. Evidently, none of these achievements would have been possible without the work of my outstanding students and postdoctoral advisees, many of whom are faculty (at Cambridge University, University of Pittsburgh, New York University, Arizona State University, National Chiao-Tong University) or are doing high caliber research in industry (Google, Tesla Inc., Samsung Corp., etc.). Advising my PhD students and guiding them to their next steps in their careers, while still being in touch with them years after their graduation, has been an absolute joy and the main reason why I joined and stayed in academia.

While the research directions that I have pursued before 2010 have addressed a set of challenging and intriguing problems, the future seemed even more challenging. Since next-generation computational substrates are likely to resemble social and biological networks and exhibit the unpredictable, complex behavior observed in these domains, I decided to look into how CAD tools can be developed or adapted to be applied to these new computational substrates, as well as their social and biological network counterparts. I have pursued these topics since 2010, in addition to continuing my work on energy-aware computing systems.

15.5.1 Computing for Life Science and Natural Phenomena

I have started working in this space about a decade ago when I commenced a collaboration with University of Pittsburgh Medical School on modeling and hardware emulation of biological networks. The medical field is, in some ways, facing the same challenges today as the computing field has encountered three decades ago when the need for increased computing capabilities required the development of tools and methodologies for system design. The astonishing pace of progress in computing was possible by the design tools, coupled with clean abstractions and predictable models encapsulating complex physical processes. My work in this space has enabled such models to be built and used in a faster than real-time analysis of biological phenomena. Starting with my initial work on hardware emulation for cell signaling networks (10^5–$10^6\times$ faster than classic simulation) and continuing to applications in cancer therapy discovery (part of the DARPA Big Mechanism program), my group and I were pleased to have our work recognized in the systems and computational biology communities and to collaborate with scientists from cancer biology, machine learning, natural language processing, and causal inference. Taking it further to other natural phenomena, we have worked on computing for sustainability, more precisely, faster than real-time forecasting of river network behavior for disaster prediction and renewable energy. Current approaches in this space do not have the accuracy or speed required for applicability in a closed loop control system, a downside which my work addresses. We have performed this work in collaboration with IBM, Penn State University, University of Texas, and University of Southern California (funded by and NSF CyberSEES grant) and were able to marry the world of automatic CAD tools with that of natural sciences.

15.5.2 Energy- and Hardware-Efficient Machine Learning

Machine learning is poised to change the landscape of computing in more ways than its broad societal applications. Indeed, hardware architectures that can efficiently run machine learning face increasing challenges due to power consumption or runtime constraints that technology, platforms, or users impose. This new direction of my research develops new modeling and optimization approaches that can enable fast and energy-efficient machine learning applications from the edge to the cloud. This started in a November 2016 afternoon when, while meeting with my students in a window-less conference room, I posed the question of power and latency characterization of ML applications. There was little work at the time on hardware accelerators for ML applications, largely from industry (e.g., TPU work from Google), but a valid question was: (1) How can we characterize ML models in terms of their hardware efficiency (power or latency); and (2) How can we co-design the ML models and the hardware so the application is served best?

From that initial meeting, my students and I have charted our path in hardware-aware machine learning. We have shown that accurate modeling of energy and runtime for neural network models running on various hardware platforms is possible (the NeuralPower framework), superseding existing methods in accuracy. We have also demonstrated that the problem of hardware-software co-design can be tackled in the context of neural network applications that are energy or runtime constrained by showing orders of magnitude improvement in state space search (the HyperPower work). Furthermore, for hardware-based neural networks, we have demonstrated that a lightweight implementation that is up to 1.5–2× more power efficient with little loss in accuracy is possible by exploiting efficient processing and storage using quantized representations (LightNNs). Finally, our work is making inroads in hardware-aware and energy-efficient image classification via fine grain labeling or layer compensated pruning, and object detection for real-time video applications using adaptive scaling. We have also demonstrated how one can find automatically in less than four hours a neural network that has highest accuracy yet satisfies hardware constraints (power or latency) on a smart phone like a Pixel 1 (in our Single Path Neural Architecture Search work). This takes the human out of the loop of neural network design and opens the chapter of truly automatic and efficient neural-hardware architecture search and co-design.

15.5.3 Beyond Academic Research

The first CRA-W/CDC Workshop on Diversity in Design Automation and Test in 2012.

Complementary to the time I spent doing research with my students, I have thoroughly relished my time teaching undergraduate and graduate students. The education component is and has been an essential ingredient in my work. I had the rewarding experience of working with more than a dozen talented CMU undergraduate students, many of which have been recognized with awards for their work under my guidance. All students I have worked with and have since graduated from CMU have either joined prestigious companies or are currently doctoral students at top universities. At the same time, I had the opportunity to interact with 60–80 students every year during course teaching at CMU. The course on energy-aware computing I have developed when I joined CMU in 2000 has been offered more than a dozen times and was still enjoying healthy enrollment and high evaluations from students. In 2018, I have developed a new course on hardware architectures for machine learning which is not only timely but also aims to bring together two areas of engineering that have largely been done separately: computer architecture and hardware design on one hand, and machine learning model development on the other hand. Last but not least, tied to my commitment of making my area and ECE in general more inclusive and welcoming to under-represented groups, I have been involved and led outreach and professional development activities at my institution and in my research community. Receiving updates from participants of these programs while they are progressing successfully through their careers lifts my spirit and confirms my hope in a better future.

15.6 Next Steps

Between 2014 and 2019, I have expanded my activities beyond scholarly research by taking two leadership roles in my department and the college. The first expands on my interest in educating our students holistically as technological leaders solving high-impact societal problems. In my position as Associate Head for Academic Affairs in ECE between 2014 and 2018, I have been overseeing all education-related activities in ECE, including the B.S. program, M.S. programs across all campuses (Pittsburgh, Silicon Valley, Rwanda, and dual degrees with other institutions), and PhD programs across all campuses (Pittsburgh, Silicon Valley, dual degrees). This was a busy time with updating our curriculum, revamping the teaching philosophy for our intro and capstone design courses, and exploring novel ways to teach and engage students in learning. I had the opportunity to work closely with my Department Head at the time, Prof. Jelena Kovacevic, who gave me *both* autonomy *and* authority to manage and lead the entire academic affairs operations of the department. It was a great time of growth for the department and for me as an academic leader.

The role I took is near and dear to my commitment to inclusion, diversity, and equity for all faculty, while providing support for faculty, from the moment they are recruited and throughout their career. As a 2014 ELATE Fellow, between 2013 and 2014, I have developed from ground up an Institution Action Project (IAP) that has established the first formal framework for faculty development at CMU, the College Engineering Center for Faculty Success (CFS), of which I have been the founding director since 2014. My Department Head at the time (Prof. Ed Schlesinger) had sponsored my ELATE participation, and the Dean of engineering at the time (Prof. Jim Garrett) had committed financial support of the center from the college. CFS is an umbrella organization that provides (1) programming and support for junior faculty through a new faculty orientation, mentoring, and workshops; (2) programs for unconscious bias awareness and mitigation (created in collaboration with Google) for both individual and group settings, such as search committees, promotion and tenure committees, graduate admissions; and (3) support for faculty development and leadership training. Not only engineering is now the only college on campus that has institutionalized support for faculty mentoring, training, development, and leadership skill building, but this has set an example for others to follow, not just within, but also outside CMU. For instance, the CFS-organized new faculty orientation in engineering has become a model for the CMU-wide new faculty orientation (started in 2016). Furthermore, CFS has run Bias Busters sessions for about 650 CMU faculty, students, and staff from across campus, of which a third are from outside engineering. Because of CFS, the vast majority of engineering faculty, and numerous staff and students, went through the Bias Busters program, putting the college at the forefront of participation across campus. Last, but not least, CFS-associated faculty are sought by other academic institutions to provide advice and feedback for their own internal programming for faculty support. I had the opportunity to participate in such activities several times with colleagues from Google and CMU at the WEPAN and Tapia conferences in 2016, the ACM/IEEE Intl. Symposium on Computer Architecture in 2018, at University of Pittsburgh in 2017 and University of Southern California in 2018. Many institutions have adopted the Bias Busters program (University of California, Berkeley; the University of Texas at Austin; Ohio State University) and are now paying it forward by educating others about the impact and ways to mitigate unconscious bias in decision-making. This was a most rewarding experience, yet truly impactful mostly for underrepresented members of the community, students, staff, and junior faculty—those who are seldom at the bottom of the power structure and mostly affected by negative effects of unconscious bias.

As I am writing this, I am weeks away from taking on the next step in my professional career: after almost two decades at Carnegie Mellon University, in December 2019 I am taking the position of Department Chair of Electrical and Computer Engineering at the University of Texas at Austin. I will cherish the two decades I have been part of the ECE community at CMU as a period of professional and personal growth, with ties and friendships that will endure—faculty, staff, students, collaborators and friends that made my job worthwhile and my work joyful. But the time is right to move on and take on a new challenge. I am thrilled to join a department with a wonderful scholarly reputation and use my background in the service of furthering its mission. I am truly excited that in my new role, I can help to make the department the best it can be and support everyone in achieving their hopes and dreams as members of the ECE community. ECE as a discipline is at an inflection point in its unique position to solve global challenges facing by our society. ECE can truly be a discipline for all, in service of the common good, while building a community that is inclusive and values diversity in people, cultures, and opinions.

This is what we, as academics, are supposed to do, and what I will personally continue doing. Faculty are (or should be) more than numbers—be them papers, citations, research dollars, graduated students—we are educators, mentors, and coaches for our students and younger colleagues. An academic institution is as good as its students, faculty, staff, and alumni think we are—by being a welcoming place for everyone, regardless of how they look, where they come from, or where they have been educated. These common values are what bonds us, not the false choice between technological pursuits and humanism—we should and can have both.

15.7 The Hidden Helpers

Our family in 2018 at Andrei's college graduation (l-r): Diana, Cristina, Andrei, Radu, and Victor.

I would be remiss if I did not mention how all this was made possible through the support of my family, friends, and network of support. My family was a steadfast supporter of my endeavors, from my time in Romania and throughout my academic career. Of course, none of this would have been possible without my husband's and children's support. Radu and I spent more than 27 years together as partners in life. A dual career is hardly an easy path for someone to be in, but our partnership is a successful story—he is my better half, my partner in life, and my first collaborator in research. I am grateful for our family, both in US and abroad, and for their patience and support. Our oldest son, Andrei, who was born while I was still a PhD student in California has just finished college last year. He used to leave me notes on how he would like to spend more time with me rather than hear me talk about "IEEE, ACM, DAC, or ICCAD.[2]" Victor is a junior in college and, although he was born in Maryland, considers himself a true Pittsburgher (a.k.a., Yinzer for those familiar with Pittsburghese). They both spent much of their early childhood traveling around US at conferences that one or both of us attended. Cristina is the only native Pittsburgher among us and she managed to bring some balance in our boy-dominated household! Last but not least, my family abroad was not merely cheering from the distance, but were physically here many times as a network of support. During my time in College Park when I was half of a commuting dual academic career couple with a young toddler and another on the way, both my mother, Gina, and my mother-in-law, Graziela, took turns in helping us, traveling thousands of miles from their homes back in Romania. My mother perhaps spent about a third of her time with us while our children were young, since 1996 when our oldest son Andrei was born, until 2010 and later when our daughter was still a toddler. My family here and back home were always supportive of whatever I sought to achieve, something that energizes me each and every day.

[2] IEEE and ACM are the main professional societies I am affiliated with. DAC (Design Automation Conference) and ICCAD (International Conference on Computer Aided Design) are the top conferences in the EDA field.

Chapter 16
Semiconductors to Light Antennas: A Woman Engineer's Career at the Turn of the Third Millennium

Lauren Palmateer

16.1 Introduction

16.1.1 The Early Days

At an early age in the 60s, 2nd grade, in Catholic school in New Jersey I started to bring chalk to school. Everybody was drawing on the pavement when I supplied the chalk; drawings, games, hop scotch, during our school breaks. Everybody came to me for a piece of chalk and then guidance as to what to draw. I told them, just draw. Then the nuns took all the chalk away and told me I could not bring any more chalk to school. When I switched to Public school, in 4th grade, I started drawing original cartoon animals, like in the Beatles Yellow Submarine characters. They were about 4 × 6 inches and I would affix them to my desk with large clear tape. Soon I started to be requested to draw pictures for other people's desks and it got so popular that my classmates started to offer me money and commissions. Soon after, the teacher said it was not OK and I was not allowed to put my pictures on desks. Therefore, I stopped. That was the end of my professional art career.

Soon after that, in 6th grade, the teachers discovered my mathematical skills and this propelled me forward into a math and science career, which is now going on professionally since 1978.

Outside of Susan, my sister, being interested in Chemistry in high school and my mother having wanted to pursue biology, but did not, our family was not engineering or science based. We knew nothing professionally about engineering, but this was the late '60s, all was whirling with fascination. I also thought clearly about how if I went into business, I might end up trying to sell something that I did not like selling. So was launched my dedication to the pursuit of science and with my credentials behind me I went sailing on.

L. Palmateer (✉)
Subtle Energy Design, San Francisco, CA, USA

A. C. Parker, L. Lunardi (eds.), *Women in Microelectronics*, Women in Engineering and Science, https://doi.org/10.1007/978-3-030-46377-9_16

The real reason I remember I wanted to become an engineer is that I wanted to have enough money to buy a great stereo. I love music; I think portable music is the best invention ever, (so far). It was not until much later, when I saw the Veterans discharge papers for World War II that I saw my father's profession at 17 years old in 1942 was listed as a musician. He must have given up that profession to have us children.

Lauren, George (our father) and Susan Palmateer circa 1965 in New Jersey

My parents were modest, they did the New York *"thing,"* moved out of Manhattan and the boroughs after they met, married and moved to New Jersey to have children and a house. They married relatively late for those times; my mom was nearly 40 when she had me. Jobs were prolific in the '50s; my parents both chose not to finish their college education. After the war, many in their '20s wanted to have fun, to enjoy life. That is what I saw. Like many other people in the '60s, they moved from Manhattan to New Jersey and had their 2.5 family and two-car garages.

Helen Palmateer (my mother) in Bergenfield New Jersey (circa 1963)

My early days were surrounded by learning and playing. Our public schools were fantastic. My early childhood was very influenced by cartoons on TV. I loved cartoons and I love the creativity of Hollywood. There was one that I can no longer find but influenced me greatly. There was this scene and it was some story and I was totally engrossed and then all of a sudden, the camera panned out and it turned out that the entire story was done on a canvas and it was an artist who was orchestrating the entire thing. I was so shocked. It was as if I was introduced to an altered state of thinking. Reality was there, and then it turned into something completely unexpected and different. I think that inspired me a great deal.

It is funny, since as my story continues to unfold, I am still working on TVs. My engineering area of expertise is the fabrication and invention of flat screen TVs. For a living I re-engineer them; analyze them for patent infringement; and basically take them apart for engineering purposes.

I graduated 8th grade and my caption in the photo referred to a song by Johnathan Richmond and the Modern Lovers. I loved music. I would go into Manhattan with my older friends to see concerts at Madison Square Garden and Carnegie Hall. The James Gang, Patty Smith, Alice Cooper, The Kinks, and so many more. My friends were musicians.

Northern Valley Regional High School Old Tappan, New Jersey, 1978 (note: they spelled my name wrong)

16.2 Antennas and Waves

16.2.1 The Bell Labs Days

My first engineering job, at the age of 18, while I was an undergraduate in electronic engineering at Monmouth College New Jersey, was at Bell Labs for Nobel Laureate Dr. Arno Penzias. I was responsible for taking the radio astronomy data, pointing the antenna at far distant galaxies, and collecting the data using a quasi-optical system. The waves coming in from the far distant galaxies were gathered and converted into electrical signals. The frequencies were from about 20 MHz to 115 GHz at that time. Later in my career, I realized how instrumental this work was in building my foundation for how antennas work. At that time, most of the optics were separate from the electronics. Only now are they really converging in engineering systems. For me they converged at the beginning. Light, waves, and electrons were all dancing together and interacting with our world and the human body, as our human body is a wonderful and mysterious antenna.

The 15-m Holmdel Horn Antenna at Bell Telephone Laboratories in Holmdel, New Jersey. Radio astronomers Robert Wilson and Arno Penzias discovered the cosmic microwave background radiation with it, for which they were awarded the 1978 Nobel Prize in physics. Horn antennas in general are popular at UHF (300 MHz to 3 GHz) [1]. (Reprinted under the terms of a Wikimedia Commons)

16.2.2 The IBM Days

I joined the research team at IBM Research Center in Yorktown Heights, NY in 1983. I was characterizing semiconductors and superconductors [2]. It was a great time. I was working with some fantastic scientists. Every day I had lunch with some of the best scientists in the world. I loved it.

16.2.3 Cornell University

Later I received a full scholarship to Cornell University for completing a PhD in Electrical Engineering in Professor Lester Eastman's group. I was engaged in making high frequency transistors. There were amazing machines and people around. Dr. David Braddock of OSEMI Inc. provided me with a picture of a typical type of machine. These types of machines were used to grow the crystals that I then made into high frequency transistors. These machines are under very high vacuum in order to reduce impurities during crystal growth; any impurities can slow down the electrons and decrease the speed of the electronics fabricated from the grown crystal wafer. I have grown crystals on these machines, but mostly I was at the receiving end, and put in my request for a particular crystal structure and then I would fabricate the wafer into a transistor device using fine line lithography techniques.

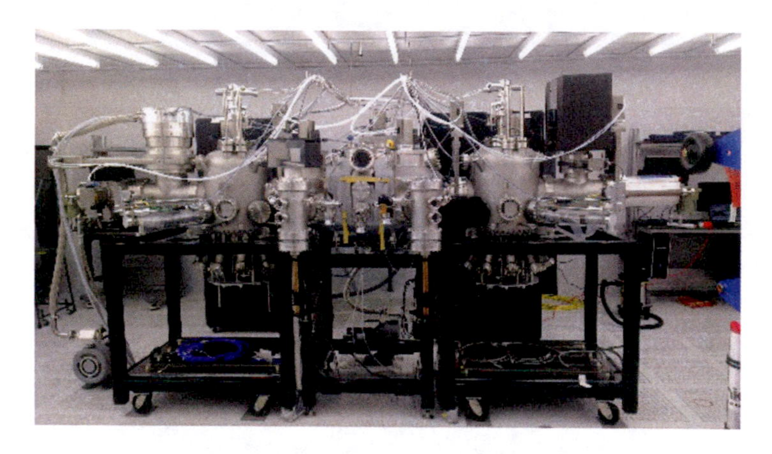

Triple growth chamber MBE (molecular beam epitaxy) cluster tool. (Photograph courtesy of Dr. David Braddock, https://www.osemi.com/)

I chose to work on Indium Phosphide (InP) and high frequency transistors for microwave devices for my PhD work. In writing a PhD, one must find something new. I show a picture of the device characteristics that revealed to me novel results for my PhD. I noticed in taking the device characteristics that the device I–V (current–voltage) curves bent up, a *"kink"* effect in the current–voltage characteristics. This was an effect of hot electrons jumping the band gap. In a review paper and in memory of Professor Lester Eastman, this device was mentioned as the record transconductance, which leads to its use in high frequency device applications [3].

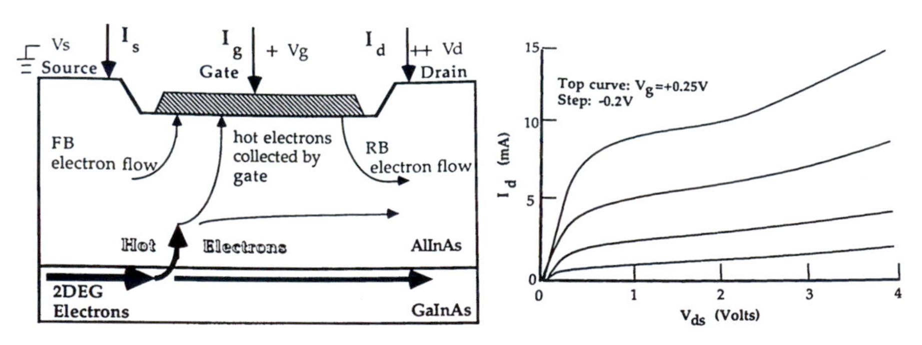

Cornell PhD Thesis 1989, MODFET (modulation doped field effect transistor). During the time of my PhD, the Apple MAC Plus had just arrived. Prior to the MAC Plus, we had an illustrator create our images for the thesis, as in the pictures shown. (Palmateer [54], Thesis, Left Figure 6.1 page 100, Right Figure 2.2 page 22)

In addition, during that time, I did probably one of the first recordings on VHS (Video Home System, Magnetic Tape Video) for the IEEE. The IEEE had just started offering their first recorded tutorials. I stood in the middle of the Bell Labs Holmdel large atrium in a black box recording my PhD dissertation.

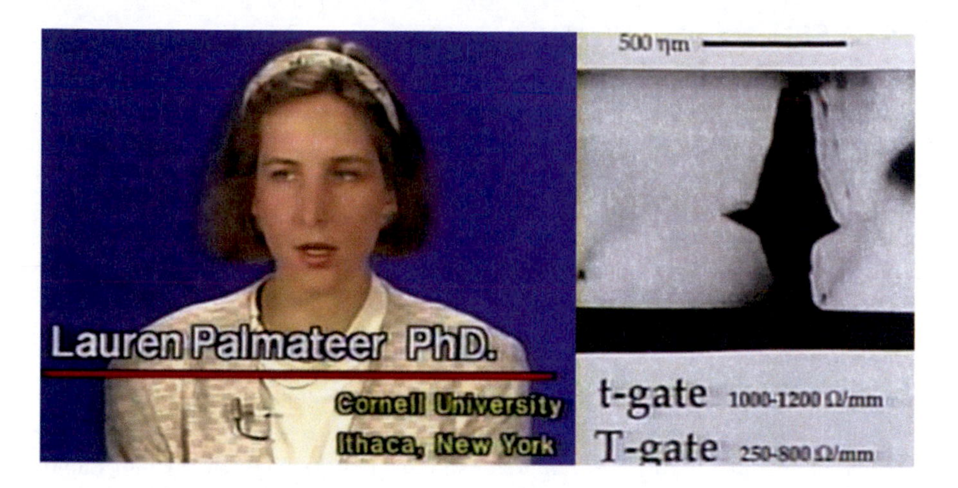

Lauren's recorded tutorial lecture in 1989 showing a submicron gate AlInAs/GaInAs MODFET (transistor) with a frequency of operation greater than 100 GHz. ([55], IEEE video recording)

Later after graduating with my PhD in Electronic Engineering, I re-joined the IBM Watson Research Lab in Yorktown Heights, NY. Soon after I joined, the GaAs project was cancelled and I was given the opportunity to join the Liquid Crystal Flat Panel Display team. This was the time in 1989 that all TVs were cathode ray tubes (CRTs). The flat panel was just being invented. We were in collaboration with Toshiba in Japan. I made my first trip to Japan to work with the Yamoto IBM Facility. It was fantastic, Japan amazed me and I returned many times after that and throughout my career: working in the Flat Panel Industry, designing and acceptance testing large area equipment in flat panel displays and attending technical conferences. I thank my bosses at IBM, Dr. Bob Wisnieff in particular, in 1990 when I was just 30 years old, I was traveling to Japan, on my own, developing and researching Flat Panel Display Technologies for IBM and visiting and working at many of the new LCD manufacturing facilities popping up in Japan at that time. I was warned that changing technical careers would be very difficult. I did not think so. The material was easy to pick up, it was exciting and I could apply and use my other areas of work and study. The field of Flat Panel Displays using Liquid Crystals was just beginning and therefore easier in many ways to have immediate impact in the technology development.

16.2.4 The French Days

After Graduating Cornell with my PhD, I headed to France for a postdoctoral position. I was working at the Observatory of Paris and Meudon, and with multiple teams at Ecole Normale Superieure and Thomson CSF Orsay Research labs. I shipped my motorcycle over too, a great 1985 Kawasaki GPZ 550, and took a grand tour of Europe.

16.3 Silicon Valley, Startups, and San Francisco

By the time that the urge to move to Silicon Valley came about in 1995, I already had a number of publications and conference presentations, some of which I have listed in the references [4–8, 53, 56]. I had traversed the world of millimeter waves in Radio Astronomy, Semiconductors and Solid-State Physics and device fabrication, Low Temperature Superconductors, and Flat Panel Display Technology and Equipment.

16.3.1 The Startup Days

As I was a member of the IBM research team that developed the first Color Laptop ThinkPad display, I was well versed in display technologies. It was in Silicon Valley I took part in Startup teams to bring various display technologies to product in both the USA and Asia. In Taiwan, I built factories to support large area glass display manufacturing processes. I developed and manufactured display technologies in rear projection television, flat panel glass X-ray replacement film technologies, and direct view MicroelectroMechanical Systems (MEMs) based e-reader display technologies. Manufacturing expertise, design of equipment, interfacing, and coordination with vendors in China, Taiwan, and Japan for the product development of the display technologies have been my area of work since 1989. This involved coordination and vendor support from a variety of international companies for parts, pieces, and processes of the manufacturing infrastructure including building large process machines for display packaging as in organic light-emitting diode display (OLED) and thin film amorphous silicon transistors (TFTs), as well as inventing new structures, thin film methods, packaging, and devices.

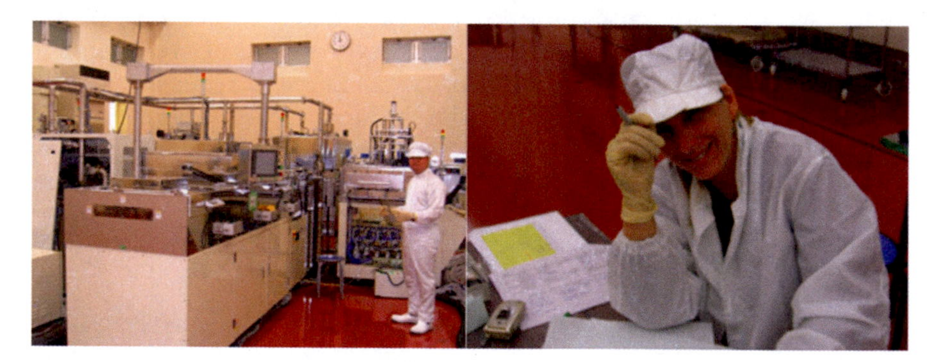

2005, Lauren aside the flat panel encapsulation machine during the acceptance testing at Japan Science Engineering Co., Ltd. (DNK) in Japan

Currently, I am a listed inventor on over 22 issued patents in the electronic display field. As a cross-disciplinary researcher, I am also embedded in other professional societies more closely associated with medical and subtle energy engineering. I am a recognized lecturer and presenter in display technologies and now presenting in interdisciplinary fields of research including novel technologies and applications in the medical and human factors areas. I have presented at ISSSEEM (International Society for the Study of Subtle Energy and Energy Medicine) and SANDS (Science and Nonduality) in 2011. I lecture in San Francisco and abroad in Paris and Brazil and was a visiting scientist in 2010 at Centro de Pesquisas Renato Archer research institution in Campinas, Brazil.

16.3.2 San Francisco Days

Living in the heart of San Francisco for over 25 years has led to intimate relationships with Silicon Valley, Startup worlds, and of course the intersection of tech and art as present in the Burning Man culture and meeting, working, and socializing with very creative and enthusiastic engineers in various fields.

16.4 Light Antennas and Metamaterials

I was told that this book chapter would help to serve others joining this field, including a short history of what I studied and how and where it led me. Therefore, now for the areas of engineering and science that I find exciting, where I think new advances are full of potential and curiosity.

Some people love to code, some people love to build factories, some people work alone, some love to work in groups. Some love to make all the money, some love to make new advances in science. Every person and every job are different. Giving advice to students or university graduates should pertain to their interests and their history and trajectory. In terms of mine, what fascinates me is the human body antenna.

How did I get involved, what piqued my interest?

I would say my background in small antenna structures and receivers in radio astronomy piqued my interest, but mostly I started to look around, and read about discoveries.

I subscribed to the magazine of photonics and biophotonics (Photonics Media). I could understand the language, as it was mostly electronics and photonics, but I started to see data that grabbed my interest and launched my fascination with neuroscience and related technologies.

One of the most intriguing articles I found relating to my interests.... I love to find anomalies... or surprising results that contradict current theories ... was light in the brains of rats. In 2017 I noticed an article by the MIT Technology Review stating that around 20 years ago, *"biologists discovered that rat brains also produce photons in certain circumstances…The light is weak and hard to detect, but neuroscientists were surprised to find it at all."* [10]

I questioned why it was thought not possible. I was always looking for the outlier, the data that was not explained easily. To me, that meant new possibilities.

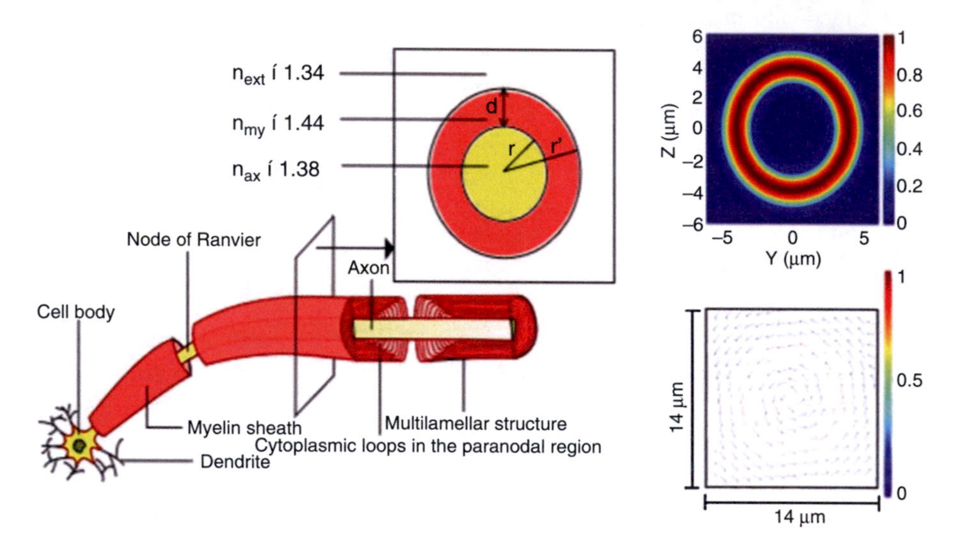

A segment of a neuron. The inner and outer radius of the myelin sheath is shown with an associated Eigen mode. As quoted in the article: ***"Finally, with the advantages optical communication provides in terms of precision and speed, it is indeed a wonder why biological evolution would not fully exploit this modality. On the other hand, if optical communication involving axons is harnessed by the brain, this would reveal a remarkable, hitherto unknown new aspect of the brain's functioning, with potential impacts on unraveling fundamental issues of neuroscience."*** **([11], p. 3 Fig. 1)**

At the same time as I was diligently submersed in the world and technology of flat panel displays, working for Qualcomm in establishing a large-scale factory in Taiwan for the MEMs reflective display technology, the Mirasol Display [12], I found my new frontier in display metrology in terms of our brainwaves. I started to realize that, in the flat panel world, not many were characterizing the effects of the displays on people biologically. It was new then, and in 2011, I gathered the data and presented results at a conference in China. I studied Neuroscience from my perspective of using brainwave technologies, specifically QEEG (Quantitative Electroencephalographic Analysis) to characterize displays. I presented and published what I considered the status in Flat Panel Display Technologies in relation to monitoring brainwaves [9].

Mini Q II Quads:
Scalp Potentials mapped to cognitive processes

a)

1. Fz Cz T3 T4 – Memory / Planning
2. F3 F4 O1 O2 – Seeing / Planning
3. C3 C4 F7 F8 – Doing / Expressing
4. P3 P4 T5 T6 – Perception / Understanding
5. Fp1 Fp2 Pz Oz – Attention / Perception

5a. T3 T4 Pz Oz – Memory / Perception

6. O1 O2 C3 C4 – Seeing / Doing
7. F7 F8 F3 F4 – Planning / Expressing
8. T5 T6 Fz Cz – Understanding / Doing

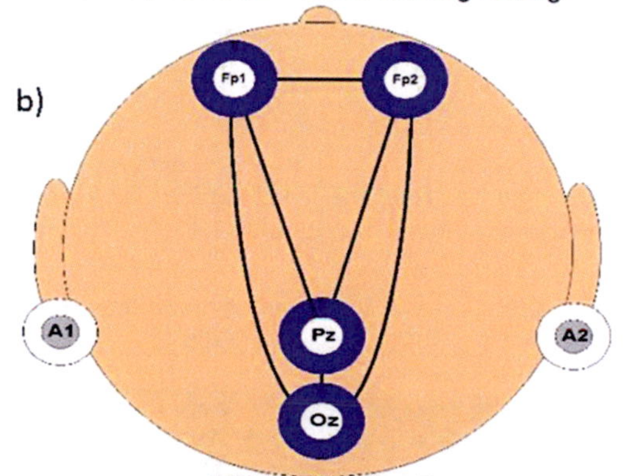

MINI-Q Position 5
"Attending and Perceiving" "The Observer; The Owl"
This position provides a primary window to logical and emotional attention, perception, and visual processing.

(a) Scalp potentials mapped to cognitive processes and (b) example of mapping to Perception ([9, 14]. China Display/Asia Display Conference 2011 Time: November 6–9 Venue: Kunshan Hotel)

It was a little too premature for me to gain traction in the industry; the article really had nothing new, just a review of the status of using EEG (electroencephalograms) for characterizing displays, to motivate and see if I could raise interest or funding. Seven years later, Dolby Labs studied these effects as I had summarized in the paper, and created a research lab in San Francisco and published the results [13]. The Dolby group has begun to collect data. These data are not easy to acquire or make conclusions; many studies that involve the brainwave technologies and intricate conclusions of the functioning of the brain are difficult. The conclusion they reached in the paper states: *"Together, these results demonstrate that properties beyond effects generated by low-level processing (e.g., visibility and brightness) can be measured with EEG. These data support the existence of signals that correlate well to modulation of viewer engagement and attention by visual stimuli."*

I became fascinated with the "BrainMaster" invented by Collura [14]. I took their classes, looked for funding, and wrote proposals for research projects both in the USA and in Brazil. It was again too soon; it was not common to have neuroscientists collaborating in the same lab as electrical engineers working on displays.

I read ferociously and attended conferences in multiple fields of disciplines. I read everything I could get my hands on in this field of how the human body could relate to all the work I had studied in semiconductors, tunneling of electrons, phonons, plasmons... looking for answers. I was intrigued and fascinated by biophotons.

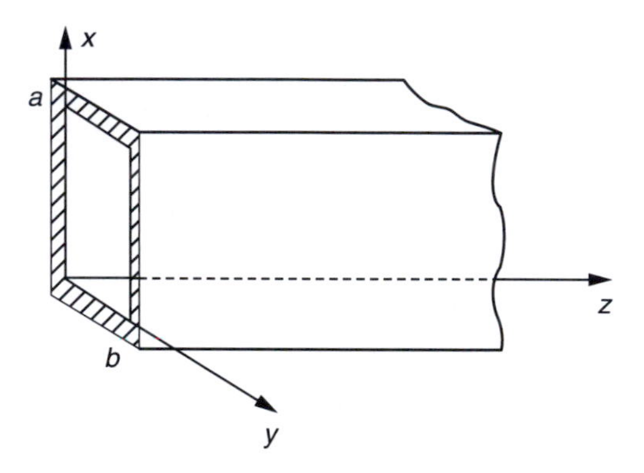

Schematic representation of an electromagnetic waveguide. The inner hollow of the brain's microtubules behaves, as it is a confined "waveguide". ([15], p. 201, Fig. 6)

The picture of the microtubular type waveguides in the brain reminded me of exactly the high frequency receivers I was working with at Bell Labs and L'Observatoire de Paris.

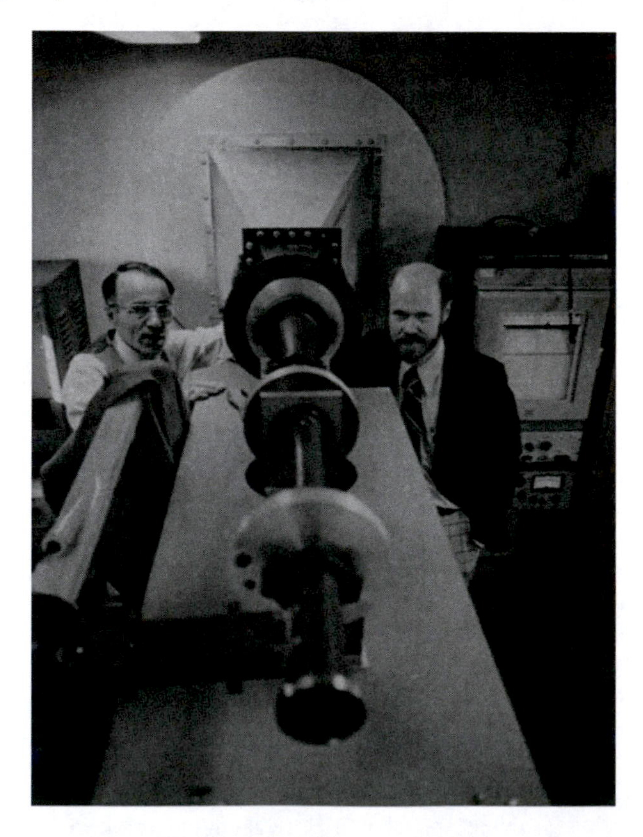

Bell Labs Scientists Robert Wilson and Arno Penzias inside the 15 m Horn Antenna at Crawford Hill, NJ. Lauren operated the antenna to gather microwaves from the stars in 1982 [16]

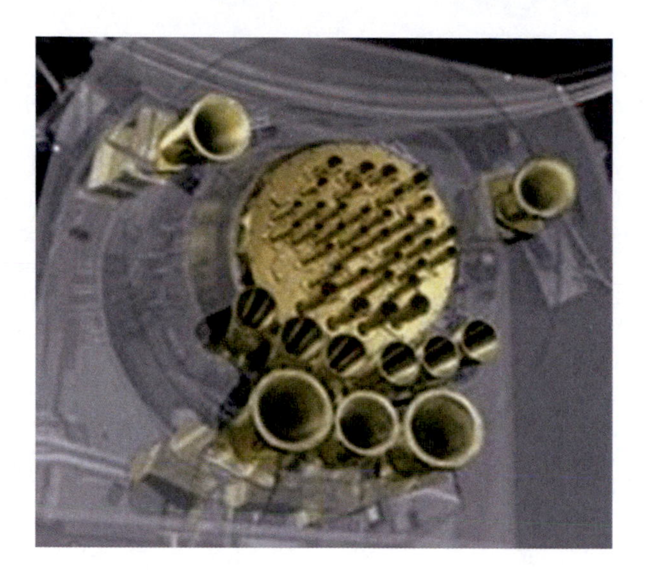

GEOStar focal plane: sub-millimeter heterodyne receiver arrays used in GEOStar. Photo courtesy of Dr. Gerard Beaudin, L'Observatoire de Paris, France, 2017

I continued to track the industries of biophotonics and neuroscience and piece the information together.

An article (referring to the data surprisingly showing light emitted in the brains of rats earlier in 2005) is quoted: *"Since then, the evidence has grown. So-called biophotons seem to be produced naturally in the brain and elsewhere by the decay of certain electronically excited molecular species"* [10]

"Ultraweak photons which are spontaneously emitted from a living body may be applicable as a non-invasive tool to characterize the physiological state of the living body" [17]. In this, same article the authors state: *"it was suggested that the central nervous system would be a good subject to study both photon communication/regulation and microtubular photon guidance"* [17].

As I continued to pursue this field, I found that bones being semiconductors was a relatively recent concept introduced by Dr. Robert Becker; the idea was at first ridiculed and rejected, and now accepted as fact [18].

Jack Kruse, MD, is a practicing neurosurgeon and a "theoretical quantum biologist." Paraphrasing the work on his website: *"Becker found that bones were semiconductors in the 1950s and 60s. This was a shocking new finding, so new that no one paid attention to this new finding. Before Becker, there were two types of current conduction ideas permeating the theories of how electric currents can run through the body, metallic and ionic. Metallic currents were the domain of electricians and solid-state physics. Ionic currents are based on large ions moving current across membranes and well-studied in cells, but suffered from short duration and poor propagation distances in cells to be used for signal transduction especially in the central nervous system. The third type was semiconduction which allowed for mas-*

sive changes in conduction with very small changes in physics and it was discovered in labs in the 1930s but few biologists understood how it worked because it works on quantum field theory which was felt to be too ridiculously difficult to be used in cells" [19].

These may at first appear to be unrelated technical advances in diverse fields, but to me, I saw it as applying my knowledge of solid-state physics to the human body.

As I continued to follow the literature from 2007 to the current day, what I noticed was an explosion in the use of metamaterials and investigations into their properties, bridging an unknown world between the human body and quantum mechanics. Since 2005, I began to see an explosion of articles in the area of *"metamaterials."* For example, scholarly publications rising to 70,000 compared to 5000 of earlier years.

Search term: "metamaterials"	Google scholar results	Google patent results
Years 1900–2005	4990	126
Years 2006–2019	701,000	6356

Table showing the large increase in work on metamaterials

The articles and patents were addressing light antennas, quantum effects, metamaterials, carbon nanotubes, and waveguide-like structures that approach the wavelength of light in size. And surprisingly, I found more work addressing ideas around room temperature and other macroscopic physics when confining electrons and photons in small waveguides, meaning these quantum effects were being seen, not as previously only at low temperature and/or high magnetic fields, but at room temperature and in biological systems.

> …In the inner hollow of brain MT (microtubes), the coherent electromagnetic field generated in water behaves as it would be *"confined"* in an electromagnetic waveguide or cavity and it is characterized by a photon frequency… [15].

> QED (Quantum Electrodynamics) coherence in the water inside brain's MT (microtubes).… considering electronic transitions rather than rotational energy levels in water molecules [15].

> ….in living systems…. energy could be stored in a thin two - dimensional layer placed beneath the cell membrane, that in this way would act as a biological superconducting medium, under dipolar propagating waves without thermal loss [15].

> In a previous work we have shown that in the water trapped inside brain microtubules could exist the conditions to allow a spontaneous QED quantum vacuum phase transition towards a macroscopic coherent quantum state characterized by a phased oscillation, at a rescaled frequency, between the water molecules states and an auto-generated electromagnetic field associated to a suitable electronic transition in them [15].

> On the other hand, it is already theoretical known and experimentally proven that a near perfect tunneling and amplification of evanescent electromagnetic waves is possible in a waveguide filled by a metamaterial [15].

Tunneling between energy states in semiconductor crystals is exactly what we did at Cornell. Why not have some of this type of functionality in our brain, or at least operate in part, based on similar functionality and principles?

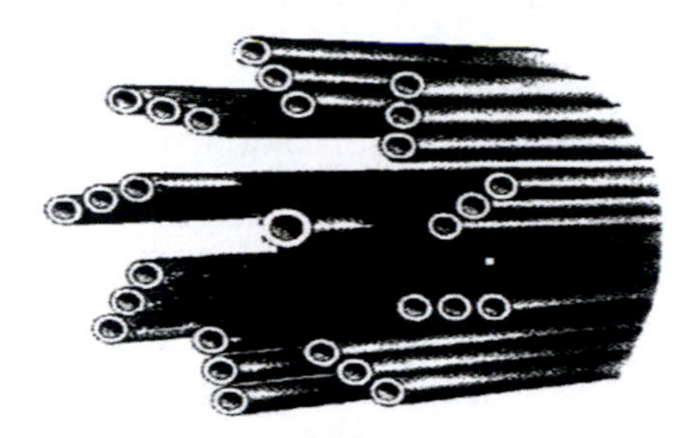

Structure of microtubule cavities in the brain. The dimensions of centrioles are close to the wavelengths of light in the infrared and visible spectrum such that they may act as phase coherent, resonant waveguides. ([20], Figure 1, page 345, Creative Commons Attribution License (CC BY 3.0))

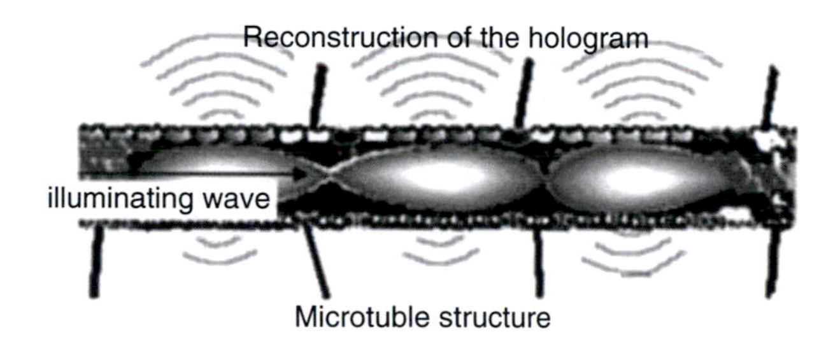

The microtubule structure in the brain is capable of storing memories as holograms. ([20], Figure 2, page 348 Creative Commons Attribution License (CC BY 3.0))

In the article, it is proposed that the substance in the microtubule cylinder has the characteristics of a metamaterial composed of sub-wavelength structures [20]. In another article... *"The amazing thing is that our antennas built themselves—we coated different classes of nanoparticles with selected sequences of DNA, combined the different families in one beaker, and nature took its course,"* Sargent said. *"The result is a beautiful new set of self-assembled materials with exciting properties"* [21].

In addition, the data showing entanglement is mounting. Another team working with Quantum Computing and measuring Three-Photon Bound States in a cloud of atoms states: *"This means these photons are not just each of them independently interacting, but they're all in together interacting strongly ... When photons go through the medium, anything that happens in the medium, they remember when they get out"* [22]. Even though these measurements were done near absolute zero, the results are telling non the less; other work in metamaterials showed occurrences of quantum entanglement at room temperature.

As my field of solid-state physics is centered around semiconductors, I also looked at discoveries there. Feng et al. [23] looked into Cavity Photon Assisted Tunneling, *"where Light photons help shuttle electrons across, a process called intra-cavity-assisted tunneling, making the device much faster."*

Would it be interesting to understand how the metamaterials and other functions of electrons and photons perhaps located in the brain may act in a similar way? Could some type of light in our brains make it work faster as the results reported for assisted tunneling? Could the neuroscientists team up with the Solid-State Physics engineers and figure out more functionality by comparing structures and results? Could the brain when properly activated in some way, maybe like a semiconductor, generate photons (biophotons) and they can do similar things as in semiconductors, and Quantum effects? What about spontaneous emission, stimulated emission, or population inversion as in a properly design laser structure? Partly this curiosity is born from the fact that it may not be so easy to experiment on the human brain directly. Looking for analogies in semiconductors that may apply to natural metamaterials in the brain can possibly give insights into neurological mechanisms and operation.

Even though the results by a research group at the Moscow Institute of Physics and Technology were obtained with a high-powered laser, still, the results are interesting and were quoted as a *"nanoantenna of sorts can scatter light in a particular direction depending on the intensity of the incident radiation... This ability is usually embedded in their geometry and the materials they are of.... The properties can be dynamically modified. When we illuminate it with a weak laser impulse, we get one results, but with a strong impulse, the outcome is completely different"* [24]. What if the intensity of the biophotons, as shown to exist in the brain of rats, is in humans and has similar functionality; its resultant output dependent on the input light intensity? Would that be interesting? Maybe, or at least some of us think so.

I also found an article in the 2015 Cornell News where electron spins were said to be controlled by high frequency sound waves, instead of where they are typically controlled by a magnetic field. The article was quoting the more detailed work of [25]. The work goes on to say, *"electron spin is a quantum phenomenon—something that happens at the atomic scale. Fuchs' group demonstrates control of a quantum phenomenon using classical vibrations. It's like reaching through a portal between two branches of physics....We're coherently interacting with this quantum thing, this spin, with something that's big and mechanical, a thing you can see with your naked eye, and that actually vibrates.... They not only created spin transitions with sound, but they also used sound to coherently control the quantum state of the spin"* [25]. Electron spins can typically be controlled by applying a magnetic field to flip the spins up or down .... However, they use sound waves .... the growing field of spintronics.

Even other phenomena involving phonons may be playing their part: such as transfer of information by converting light to sound waves as is done in electronic devices. *"...phonons can function as unique links between radiofrequency and optical signals, allowing access to quantum regimes"* [26].

Who is to say this is not happening in our brains? We have biophotons shown to exist in rat brains and in the human body and we have acoustic waves through our ears or elsewhere, why not phonons in our brains with then enabling access to quantum regimes?

New data seems to be popping up all the time, as in 2017 an article showed *"Scientists have demonstrated entanglement between 16 million atoms in a crystal crossed by a single photon; reinforcing the quantum theory that entanglement can persist in macroscopic physical systems"* [27]. *"Macroscopic systems"* is the intriguing part, as it is not just a single isolated small system.

As one bridges the work in microelectronics to work in the neurosciences, I quote a few of the observations; I am sure there are many, but these are the ones I chose to highlight.

A simple look at Wikipedia for *"Optogenetics"* says: *"In 2010, optogenetics was chosen as the 'Method of the Year' and across all fields of science and engineering by the interdisciplinary research* [28–30] *and 'Breakthroughs of the Decade'"* [31]. Optogentics plays its part; it is shown that a silicon-based neural implant can control the electrical activity of brain cells by shining multicolor light into the brains of awake mice [30].

In 2010, Karl Deisseroth at Stanford University was awarded *"for his pioneering work on the development of optogenetic methods for studying the function of neuronal networks underlying behavior"* [32], when in 1999 it was considered *"far-fetched"*—the possibility of using light for selectively controlling precise neural activity [33].

And then of course there is the fascinating subject of using novel electromagnetic brain stimulation techniques in treating psychiatric diseases [34].

16.5 Neuroscience and the Human Body Antenna

Neuroscience had been claimed by many as the new *"golden age."* I was so happy to hear that this series of the Springer *"Women in Microelectronics,"* is being combined with work in the frontiers in neuroscience. Neuroscience, the mind and its mysteries, are what now, after a long career in electronics, is what I find the most interesting and reveals groundbreaking work. It bridges into the world of electronics, with new and current research showing, for example, that metamaterials are present in biological systems, and metamaterials are able to act as tiny antennas and display other observed anomalies that lead to Quantum Electrodynamics (QED) systems. In pursuit of a basic understanding of this field, I turned to other electrical engineering personalities who have followed this same path of interest.

Federico Faggin is one such engineer, making the bridge from the *"golden age of Microelectronics" to the "golden age of Neuroscience."* He was born in the 1940s and is an Italian physicist, inventor, and entrepreneur, widely known for designing the first commercial microprocessor. Frederico is extensively published in this area and speaks frequently on this subject.

> *Science cannot explain the fact that we are conscious. Therefore, we must consider the possibility that consciousness may be an irreducible property of nature, in which case consciousness must be present in the elementary particles of which all matter is built. This idea is generally considered crazy. That's why scientists give no reality to the inner world which is made of sensations, feelings, emotions, and thoughts, which are all about consciousness, and give only value to what is external. .... Can we somehow guide our conclusions into this kind of realization? Yes, we can. We can by wanting to know and by looking at the right places, through meditation, for example. Through questioning yourself little by little, you will put yourself in a position to experience beyond what we ordinarily experience. Expanding one's consciousness is essential to being able to understand a broader reality."* [35]

> *"... people who recognize that science cannot explain the nature of awareness are on the rise all over the world."* [36]

Diane Powell MD is another maverick and spokesperson in this field. Her work explores the unexplained connection in consciousness between twins. Her current research focuses on autistic savants as she can observe reproducible results. The work opens the door to the study of exotic physics in non-closed systems [37].

When we talk about the brain and the heart relationship, one of my favorite successful institutions researching this is HeartMath. Dr. Rollin McCraty at HeartMath speaks of the *"heart-brain connection"*: *"Most of us have been taught in school that the heart is constantly responding to 'orders' sent by the brain in the form of neural signals. However, it is not as commonly known that the heart actually sends more signals to the brain than the brain sends to the heart! Moreover, these heart signals have a significant effect on brain function—influencing emotional processing as well as higher cognitive faculties such as attention, perception, memory, and problem-solving. In other words, not only does the heart respond to the brain, but the brain continuously responds to the heart"* [38].

It reminds me of when I discovered the work of Dr. Candice Pert—Molecules of Emotion [39]. She showed that the same molecules that are linked to emotional states in the brain are the same as located in organs and other tissues of the human body. Literally, our bodies speak the same language as our brains [40].

Viewing the human body as a great electrical and chemical antenna, receiver, and transmitter, I find it interesting to look at the fields of the heart. From a very simple web search on *"Which is stronger the heart or the brain?"* One finds: *"The heart emits more electrical activity than the brain. The heart emits an electrical field 60 times greater in amplitude than the activity in the brain and an electromagnetic field*

5,000 times stronger than that of the brain. The electromagnetic field of the heart is incredibly strong."

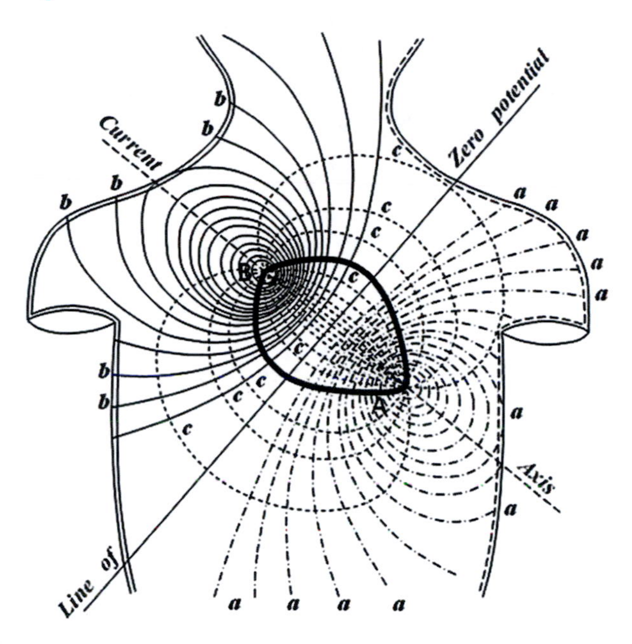

Electric field of the heart on the surface of the thorax, recorded by Augustus Waller (1887). The curves (a) and (b) represent the recorded positive and negative isopotential lines, respectively. These indicate that the heart is a dipolar source having the positive and negative poles at (A) and (B), respectively. The curves (c) represent the assumed current flow lines. ([41], p. 51 Fig. 1.17)

The amount of work both inside and outside of mainstream technical articles on this field of neuroscience and how information, the electric and the magnetic fields permeate the body, is bewildering and staggering, as seen in just one article detailing many contributions in the field [42].

With such a rich landscape of research and experiment, I can only list here a few of the people that I have studied and read their work. One such person is Dr. Shelli Joye, who I met here in San Francisco. From her book, she illustrates the eight feasible bands for consciousness and in her book, she illustrates how the microwave microtubular may carry this information [43].

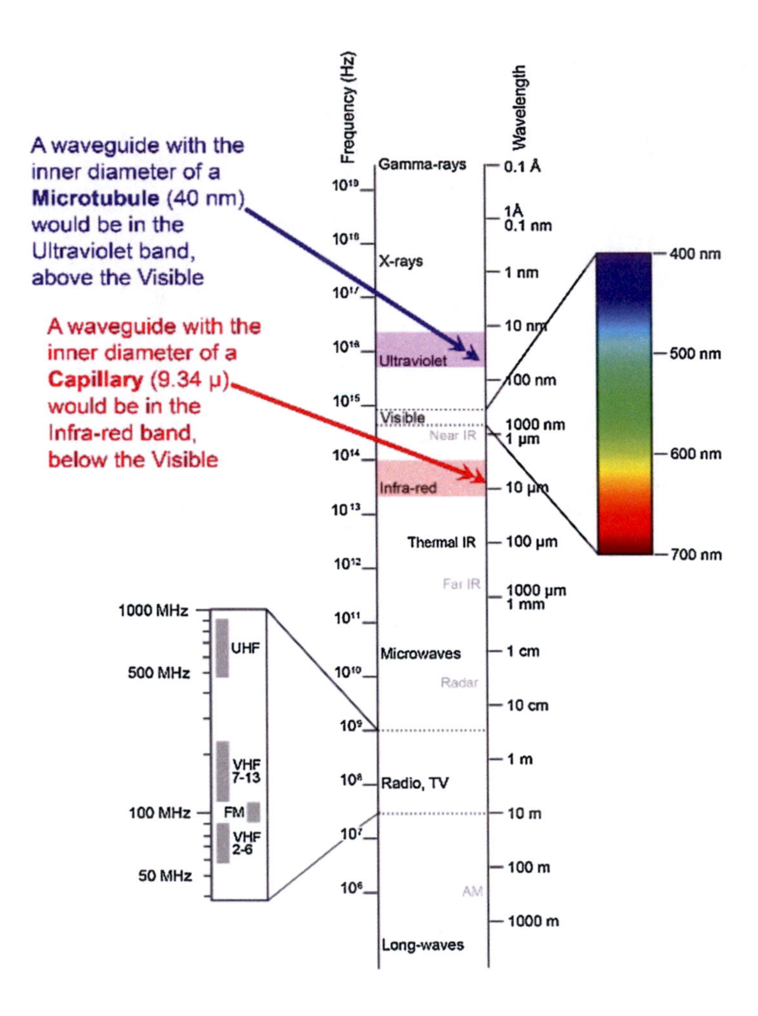

Microtubules and capillaries as waveguides ([43], p. 130. Fig. 8.3)

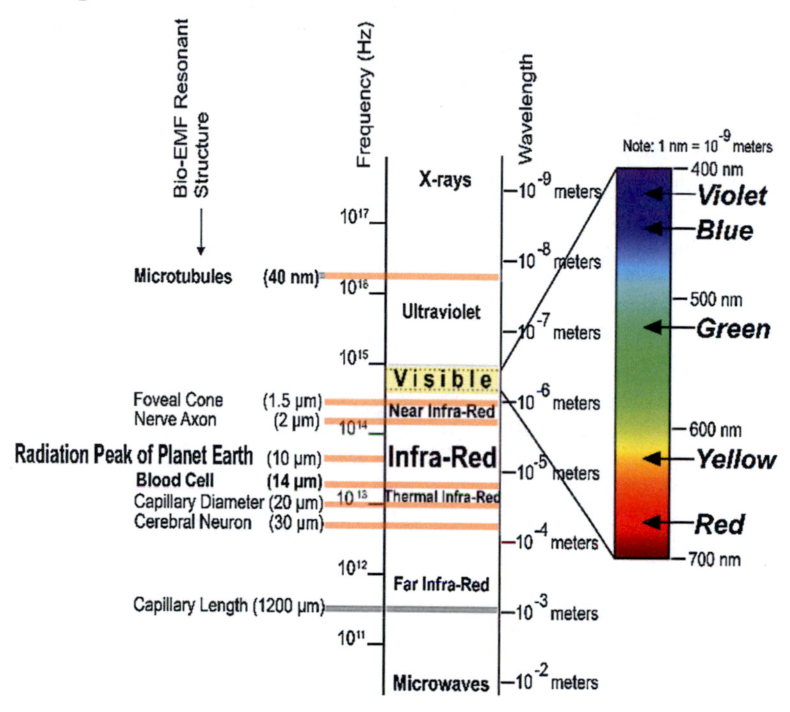

Developing supersensible perception ([43], p. 138 Fig. 8.6)

16.6 The Intellectual Property Days

It is not, as of yet, easy for an electrical engineer to make a living in the neurosciences. As for myself, having in general moved out of working in the laboratory, with many patents and having been trained in the Strategic Intellectual Property Division of Qualcomm, I work as what is termed an independent *"subject matter expert" or "patent engineer"* consultant. Prior to being an independent consultant, I worked for Rovi who merged with TiVo. At TiVo, I was taking apart TVs to determine if the Rovi/TiVo patents were infringed. In general, as a consultant, my work involves in-depth reverse engineering of complex semiconductor structures to reveal the use of a material or device in a product for asserting a patent, for licensing fees or litigation. It can involve prior art searches to determine if an idea is patentable and helping inventors determine patentable ideas. It can involve using commercial programs that synthesize complex *"big data"* (worldwide patent databases) to be analyzed and used by businesses, venture firms, corporations, and students. As the example of one of the types of analysis, the figure below shows a sample mapping from such a program, INNOGRAPHY® an Intellectual Property Intelligence Software suite of CPA Global.

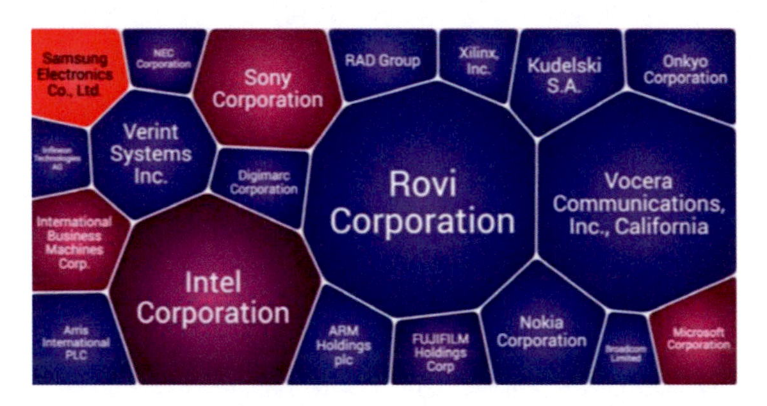

Highly-rated Google patent US7165175 (forward citations) CPA Global Innography Software PatentScape heat map visualization (visualization mapping courtesy of CPA GLOBAL) [44]

The conceptual process for utilizing the patent databases to reveal new market research and patent landscapes is detailed in a paper by Yang et al. [45]. Other papers reporting such data are in the references [46–48]. Such data can be used to determine if a technology space is an open *"white space"* as an area to pursue or the data could reveal it is highly populated in ownership by various entities and would mean licensing, mergers, or other options to be able to work in that technology area. It can also help to guide companies, attorneys, individuals, and students into areas of inventiveness as one visualizes and maps the data on keywords, technologies, and companies. It is a valuable tool for market research as well and can give insights into technology through the lens of innovation.

In addition to the patent landscaping platforms, as the changing world of startups and innovation moves, so does big data in the form of artificial intelligence-based platforms helping startups, universities, and incubators create and innovate by combining startup entrepreneurs, artificial intelligence, and data science experts [49].

16.7 Future Progression

What could I recommend to new people entering the field of electronics, if not interested in neuroscience or other topics I suggested such as quantum biology, photobiomodulation, and optogenetics?

What is the problem with cross-disciplinary fieldwork? It is needed in our world, but hard to pursue. Experts are required; when one crosses into other fields, one can be lost in an inability to dialogue. However, I believe it is needed. I also think women have an extra special talent for it, in our world today, or many do.

In general, for new professionals, I would suggest reading journals in areas of interest and go to conferences (trade shows are free or cheap for entry). In addition, doing a Google patent search or Google scholar search on keywords of interest may spur on new ideas and technology areas of interest. Also find professional societies of interest, such as Woman in 3D Printing [50].

In preparing for this book chapter, I interviewed a few women in positions I felt were progressive, thoughtful, and technically advanced.

I asked Dr. Beverly Rubik of Frontier Sciences what she would want to say to incoming professionals in the field: Her answer was: *"Continue to pose questions that you want to learn. As you want to learn about it, doing something novel will reveal itself and people will appreciate it finding that is part of the experience."* Her area of expertise is life sciences and biology related fields. Her advice was to *"elucidate the energies of the human body so we can have beneficial technologies. We are in need of building a 'human energy project' as like the human genome project—The human energy field."*

For me in particular, being able to explore Quantum Mechanics in Consciousness fascinates me, but as I indicated, this is a difficult job to support oneself on, I think, but maybe that will change in the future.

How could it not be that the environment interacts with our human body and very profoundly?

I found the solace that I was looking for in the fields of neuroscience and consciousness ... that the future of Science will encompass more of the feminine: Inward, Subtle and Powerful.

Now that so much data is available, showing and demonstrating the human body as an electrical and chemical antenna, receiving and transmitting, of course then, we would like to study the effects of electromagnetic waves in more depth on the human body, in mainstream universities and projects. Even though I have not been able to work in a job in my field of passion, I found happiness in reading about it. I have inspired audiences and given talks at conferences and universities in these areas and continue to invent, and present novel ideas for people to review and consider.

Since we really do not know much about the true workings of the brain and its relationship to consciousness, we can take the operations and functions that we know to exist in similar structures such as semiconductors and metamaterials, and at least out of curiosity, be considering that these functions may exist in our ourselves, our brain and nervous system. The most intriguing for me is that a lot of the

work with natural metamaterials indicates that those functioning properties of metamaterials could exist in the brain and display similar effects as in the external electronic systems we build, which brings out the possibility of all the beauty of Quantum Electrodynamics and that familiar Einstein's *"Spooky Action at a Distance"* taking place right here in our own bodies and minds.

One can contribute to a productive and exciting world. Good luck! Let the force be with you. Nobody ever said this would be easy. Now, if not anybody believes in the plausibility of telepathy and action at a distance, I believe quantum electrodynamics in biological systems and metamaterials is telling us otherwise, look at the science and have fun.

In a 1935 paper, Einstein and his co-authors showed how entanglement leads to what is now called quantum nonlocality, the "eerie" link that appears to exist between entangled particles. Einstein described quantum mechanics as "spooky" because of the instantaneousness of the apparent remote interaction between two entangled particles. *"If two quantum systems meet and then separate, even across a distance of thousands of light-years, it becomes impossible to measure the features of one system (such as its position, momentum and polarity) without instantly steering the other into a corresponding state"* [51].

It is almost 100 years since Einstein published his results of *"action at a distance"* and other strange phenomena, Hollywood has accepted it; when will we (see patent US6506148)?

Original art of Lauren's fingerprint. Philippe Jestin, Artist www.philippejestin.com. Resin on wood panel [52]

Thank-you to all my teachers and my gratitude to the people having initiated, written and edited this book, their dedication, and to the people reading the book and their quests in looking for new answers and curiosities yet to be discovered.

References

1. The Holmdel Horn Antenna (1962). Available at https://en.wikipedia.org/wiki/Holmdel_Horn_Antenna. Accessed 15 September 2019
2. Kirtley JR, Schlesinger Z, Theis TN, Milliken FP, Wright SL, Palmateer LF (1986) Voltage-controlled dissipation in the quantum Hall effect in a laterally constricted two-dimensional electron gas. Phys Rev B 34:5411

3. Yoder MN (2002) 30 years of accomplishments in compound semiconductor materials and devices attributable to Prof. Lester F. Eastman. In: IEEE Lester Eastman conference on high performance devices. IEEE, Piscataway, pp 34–39
4. Palmateer LF, Tasker PJ, Schaff WJ, Nguyen LD, Eastman LF (1989) dc and rf measurements of the Kink Effect in 0.2 micrometer Gate Length AlInAs/GaInAs/InP MODFETS. Appl Phys Lett 52(21):2139
5. Palmateer LF, Tasker PJ, Itoh T, Brown AS, Eastman LF (1986) Microwave characterization of 1 micron Gate Al0.48In 0.5 2 As/Ga 0.4 7 In 0.5 3As modulation doped field effect transistors. In: WOCSEMMAD (workshop on compound semiconductor materials & devices) conference, San Francisco, CA, February 10–12, 1986
6. Palmateer LF, Tasker PJ, Itoh T, Brown AS, Griem T, Poli L, Schaff WJ, Wicks GW, Eastman LF (1986) Characterization of Al 0.4 8 In 0.5 2 As/Ga 0.4 7 In 0.53 As/InP MODFETs and a Comparison to AlxG a 1 - xAs/GaAs and Strained Layer A10.1 5Ga 0.8 5As/Ga 0.8 5 In 0.1 5As MODFETs. NATO. In: 3rd InP NATO Workshop, Cape Cod, MA September 22–25, 1986
7. Palmateer LF, Tasker PJ, Itoh T, Brown AS, Wicks GW, Eastman LF (1987) Microwave characterisation of 1 μm-gate Al0.48In0.52As/Ga0.47In0.53As/InP MODFETs. Electron Lett, 23, n 1, p 53
8. Palmateer LF, Tasker PJ, Lepore LD, Eastman LF (1988) Observation of excess gate current due to hot electrons in 0.2 micron gate length 100GHz f(t) AlInAs/GaInAs/InP MODFETS. In: Proceeding of the 1988 international symposium on gallium arsenide and related compounds, Atlanta Georgia 1988 (Inst. Phys. Ser. Conf.). Available at https://catalog.princeton.edu/catalog/SCSB-2770460. Accessed 16 January, 2020
9. Palmateer LF, Acosta-Urquidi J, Pellegrini Mammana V, den Engelsen D, Williams J (2011) Characterization of electronic displays: current methods to human centered approaches as EEG brainwave monitoring. In: China display/Asia display conference 2011. November 6-9 Kunshan Hotel, Kunshan, China
10. MIT Technology Review (2017) Emerging technology from the arXiv. Are there optical communication channels in the brain? arxiv.org/abs/1708.08887. Available at https://www.technologyreview.com/s/608797/are-there-optical-communication-channels-in-our-brains/. Accessed 16 January, 2020
11. Zarkeshian P, Kumar S, Tuszynski J, Barclay P, Simon C (2017) Are there optical communication channels in the brain? arXiv:1708.08887[physics.bio-ph]. Available at https://arxiv.org/abs/1708.08887. Accessed 16 January, 2020
12. Kozlowski M (2018) GoodReader "the rise and fall of qualcomm mirasol e-readers". Available at https://goodereader.com/blog/electronic-readers/the-rise-and-fall-of-qualcomm-mirasol-e-readers. Accessed 16 January, 2020
13. Darcy D, Gitterman E, Brandmeyer A, Daly S, Crum P (2016) Physiological capture of augmented viewing states: objective measures of high-dynamic-range and wide-color-gamut viewing experiences. Electron Imaging Human Vis Electron Imaging 16:1–9. https://doi.org/10.2352/ISSN.2470-1173.2016.16.HVEI-126
14. Collura T (2015). https://www.brainmaster.com/#nav-two. Accessed 16 January, 2020
15. Caligiuri LM, Musha T (2015) Superradiant coherent photons and hypercomputation in brain microtubules considered as metamaterials. Int J Circuits Syst Signal Process 9:192–204
16. Matthews C (2015) Fortune "The reincarnation of bell labs". Available at https://fortune.com/2015/02/02/bell-labs-real-estate-revival/. Accessed 16 January, 2020
17. Yoon YZ, Kim J, Lee BC, Kim YU (2005) Changes in ultraweak photon emission and heart rate variability of epinephrine-injected rats. Gen Physiol Biophys 24:147–159
18. Becker R (1985) The body electric. Electromagnetism and the foundation of life. Morrow, New York
19. Kruse J (2013) EMF 8: quantum bone. Available at https://jackkruse.com/emf-8-quantum-bone/. Accessed 16 January, 2020
20. Musha T (2012) Holographic view of the brain memory mechanism based on evanescent. Superluminal Photons 3(3):344–350

21. Tikhomirov G, Hoogland S, Lee PE, Fischer A, Sargent EH, Kelle SO (2011) DNA-based programming of quantum dot valency, self-assembly and luminescence. Nat Nanotechnol 6: 485–490
22. Liang QY, Venkatramani AV, Cantu SH, Nicholson TL, Gullans MJ, Gorshkov AV, Thompson JD, Chin C, Lukin MD, Vuletić V (2018) Observation of three-photon bound states in a quantum nonlinear medium. Science 359(6377):783–786. https://doi.org/10.1126/science.aao7293
23. Feng M, Qiu J, Wang CY, Holonyak N Jr (2016) Tunneling modulation of a quantum-well transistor laser. J Appl Phys 120:204501
24. Baranov DG, Makarov SV, Krasnok AE, Belov PA, Alù A (2016) Tuning of near- and far-field properties of all-dielectric dimer nanoantennas via ultrafast electron-hole plasma photoexcitation. Laser Photonics Rev 10(6):1009–1015
25. MacQuarrie ER, Gosavi TA, Moehle AM, Jungwirth NR, Bhave SA, Fuchs GD (2015) Coherent control of a nitrogen-vacancy center spin ensemble with a diamond mechanical resonator. Optica 2(3):233–238
26. Merklein M, Stiller B, Vu K, Madden SJ, Eggleton BJ (2017) A chip-integrated coherent photonic-phononic memory. Nat Commun 8:574
27. Fröwis F, Strassmann PC, Tiranov A, Gut C, Lavoie J, Brunner N, Bussières F, Afzelius M, Gisin N (2017) Experimental certification of millions of genuinely entangled atoms in a solid. Nat Commun 8:907. https://doi.org/10.1038/s41467-017-00898-6
28. Deisseroth K (2011) Nature methods. Optogenetics. https://doi.org/10.1038/nmeth.f.324
29. Nature Methods, Editorial (2011) Method of the year 2010. Available at https://www.nature.com/articles/nmeth.f.321. Accessed 16 January, 2020
30. Pastrana E (2011) Primer on optogenetics: optogenetics: controlling cell function with light. Nat Methods 8(1):24–25. https://doi.org/10.1038/nmeth.f.323
31. Science (2010) Stepping away from the trees for a look at the forest. Science 330(6011):1612–1613. https://doi.org/10.1126/science.330.6011.1612
32. Deisseroth K (2010) Scientific American optogenetics: controlling the brain with light. Available at https://www.scientificamerican.com/article/optogenetics-controlling/. Accessed 16 January, 2020
33. Crick F (1999) The impact of molecular biology on neuroscience. Philos Trans R Soc B 354(1392):2021–2025. https://doi.org/10.1098/rstb.1999.0541
34. Deisseroth K (2010) hfsp (human science frontier program. https://www.hfsp.org/hfsp-nakasone-award/2010-karl-deisseroth. Accessed 16 January, 2020
35. Accardi C (2019) L'Italo-Americano. "Federico Faggin – The man behind the genius". Available at https://italoamericano.org/story/2019-2-1/federico-faggin
36. We the Italians (2019) Federico Faggin (creator of the first microchip). Available at https://www.wetheitalians.com/interviews/italian-america-who-changed-world-meet-great-federico-faggin. Accessed 16 January, 2020
37. Powell DH (2009) The ESP enigma: the scientific case for psychic phenomena. Walker Publishing Company, New York City
38. McCraty R (2019). Available at https://www.heartmath.com/science/. Accessed 16 January, 2020
39. Pert CB (2010) Molecules of emotion: the science behind mind-body medicine. Simon and Schuster, New York City
40. The Telegraph (2013). Candace Pert – obituary [online]. Available at https://www.telegraph.co.uk/news/obituaries/medicine-obituaries/10486776/Candace-Pert-obituary.html. Accessed 16 January, 2020
41. Malmivuo J, Plonsey R (1995) Bioelectromagnetism: principles and applications of bioelectric and biomagnetic fields. Oxford University Press, New York
42. Meijer DKF, Geesink HJH (2017) Consciousness in the universe is scale invariant and implies an event horizon of the human brain. NeuroQuantology 15(3):41–79. https://doi.org/10.14704/nq.2017.15.3.1079
43. Joye SR (2019) Developing supersensible perception. Inner Traditions, Rochester

44. CPA Global (2019) Geek out guide competitive intelligence. Available at https://learn.cpaglobal.com/hubfs/2019%20CPANA/Nurture%20Campaign/Competitive%20Intelligence/2019%20Competitive%20Intelligence%20Geek%20Out%20Guide%20FlipBook/index.html?page=1. Accessed 16 January, 2020
45. Yang X, Yu X, Liu X (2018) Obtaining a sustainable competitive advantage from patent information: a patent analysis of the graphene industry. Sustainability 10(12):4800
46. Ding Q, Luo R, Tong Q, He H, Zhong J (2017) Patents analysis on magnetic resonance imaging and data processing technology. In: 2nd IET International Conference on Biomedical Image and Signal Processing, pp. 1–5. Accessed at https://digital-library.theiet.org/content/conferences/10.1049/cp.2017.0101. Accessed 16 January, 2020
47. Liu X, Yang X (2019) Identifying technological innovation capability of high-speed rail industry based on patent analysis. In: 2019 8th international conference on industrial technology and management (ICITM). IEEE, Piscataway, pp 127–131. https://doi.org/10.1109/ICITM.2019.8710710
48. Qiu H, Yang J (2018) An assessment of technological innovation capabilities of carbon capture and storage technology based on patent analysis: a comparative study between China and the United States. Sustainability 10(3):877
49. Beddows J (2019). https://www.decisionplatform.io/. Accessed 16 January, 2020
50. Toure N (2019). https://womenin3dprinting.com/about-us/. Accessed 16 January, 2020
51. Crull E (2019) If you thought quantum mechanics was weird, you need to check out entangled time. Available at https://www.sciencealert.com/if-you-thought-quantum-mechanics-was-weird-wait-til-you-check-out-entangled-time. Accessed 16 January, 2020
52. Jestin P (2016) Four to trace her. [Resin on wood panel 20″ x 20″ x 1.5″] (Philippe Jestin, Artist's Collection for purchase). www.philippejestin.com. Accessed 16 January, 2020
53. Palmateer LF (1987) Characterization of submicron AlInAs/GaInAs/InP MODFETS. In: WOCSEMMAD (workshop on compound semiconductor materials & devices) conference, Hilton Head, South Carolina, March 1987
54. Palmateer LF (1989) DC and RF characterization of gallium-indium-arsenide/aluminum-indium-arsenide/indium-phosphide modulation doped field effect transistors for millimeter wave device applications PhD. Cornell University. Available at https://cornell.worldcat.org/title/dc-and-rf-characterization-of-gallium-indium-arsenidealuminum-indium-arsenideindium-phosphide-modulation-doped-field-effect-transistors-for-millimeter-wave-device-applications/oclc/703489928?referer=di&ht=edition. Accessed 16 January, 2020
55. Palmateer LF (1989) IEEE video recording, DC and RF characterization of MODFETs for millimeter wave device applications. IEEE, Piscataway
56. Palmateer LF (1994) Measurements evaluating photosensitive TFT off currents in AMLCDs under display operating conditions. In: 1994 SID international symposium digest of technical papers, pp 441–443

Index

A. C. Parker, L. Lunardi (eds.), *Women in Microelectronics*, Women in Engineering and Science, https://doi.org/10.1007/978-3-030-46377-9

O

P

Q

R

S